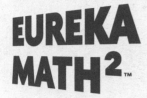

EUREKA MATH²™

D0989262

A Story of Units®

Fractional Units ▸ 4

APPLY

Great Minds® is the creator of *Eureka Math*®,
Wit & Wisdom®, *Alexandria Plan*™, and *PhD Science*®.

Published by Great Minds PBC.
greatminds.org

Copyright © 2021 Great Minds PBC. All rights reserved. No part of this work may be reproduced or used in any form or by any means—graphic, electronic, or mechanical, including photocopying or information storage and retrieval systems—without written permission from the copyright holder.

Printed in the USA

1 2 3 4 5 6 7 8 9 10   LSC   25 24 23 22 21

ISBN 978-1-64497-657-9

Contents

Foundations for Fraction Operations

Copyright © Great Minds PBC

FAMILY MATH
Fraction Decomposition and Equivalence

Dear Family,

Your student is learning that fractions, like whole numbers, can be decomposed into sums of parts. For example, the whole number 4 is the sum of $1 + 3$ and the fraction $\frac{4}{5}$ is the sum of $\frac{1}{5} + \frac{3}{5}$. They break apart fractions and write addition equations with the parts. They use models to break apart whole numbers and fractions and see that fractions can be broken apart in many ways. They combine whole numbers with fractions less than 1 to write mixed numbers. Your student also learns to rename a mixed number as a fraction. Renaming fractions with models and equations prepares your student for adding and subtracting fractions in future lessons.

Key Term

mixed number

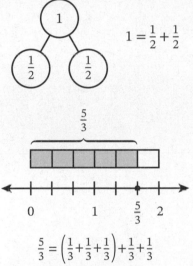

$$1 = \frac{1}{2} + \frac{1}{2}$$

$$\frac{5}{3} = \left(\frac{1}{3} + \frac{1}{3} + \frac{1}{3}\right) + \frac{1}{3} + \frac{1}{3}$$

Students use familiar models, such as number bonds and tape diagrams, to break apart numbers into the sums of fractions.

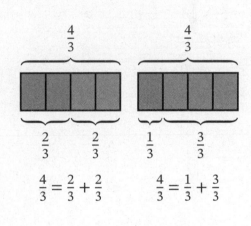

$$\frac{4}{3} = \frac{2}{3} + \frac{2}{3} \qquad \frac{4}{3} = \frac{1}{3} + \frac{3}{3}$$

Students use tape diagrams to see how fractions can be broken apart in different ways.

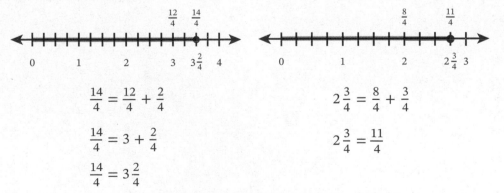

$$\frac{14}{4} = \frac{12}{4} + \frac{2}{4}$$

$$\frac{14}{4} = 3 + \frac{2}{4}$$

$$\frac{14}{4} = 3\frac{2}{4}$$

$$2\frac{3}{4} = \frac{8}{4} + \frac{3}{4}$$

$$2\frac{3}{4} = \frac{11}{4}$$

Students use number lines to support their understanding when writing a fraction as an equivalent mixed number or when writing a mixed number as an equivalent fraction.

Copyright © Great Minds PBC

At-Home Activity

Is This Share Equal?

Help your student practice breaking apart fractions in different ways. Start with a food item that is already divided into equal parts, such as a chocolate bar. Another option is to cut a food item, such as a piece of bread or fruit, into an even number of equal parts. Ask your student what fraction describes 1 part of the whole. Ask them what fraction of the whole each of you would get if you shared the whole equally. Then discuss an example of a share that is not equal such as the following example. This example is for a chocolate bar that has 12 sections.

- The fraction that describes 1 part would be $\frac{1}{12}$.

- To share equally between two people, each person gets $\frac{6}{12}$ of the bar. Discuss all the different ways you could decompose $\frac{6}{12}$ such as $\frac{6}{12} = \frac{1}{12} + \frac{1}{12} + \frac{1}{12} + \frac{1}{12} + \frac{1}{12} + \frac{1}{12}$ or $\frac{6}{12} = \frac{2}{12} + \frac{4}{12}$ and so on. Ask whether there is a way to share $\frac{6}{12}$ equally such as $\frac{6}{12} = \frac{3}{12} + \frac{3}{12}$.

- A share that is not equal would happen when one person gets $\frac{3}{12}$ of the whole bar and the other person gets $\frac{9}{12}$.

Name _____ Date _____

1. Complete the equations to match the partitioned circle. The circle represents 1.

3 thirds = ___1___ third + ___1___ third + ___1___ third

$$\frac{3}{3} = \frac{1}{3} + \frac{1}{3} + \frac{1}{3}$$

I know the whole circle represents 1.

The whole circle is partitioned into 3 equal parts. Each part is 1 third, or $\frac{1}{3}$.

I can complete the equations in unit form and standard form.

3 thirds = 1 third + 1 third + 1 third

$$\frac{3}{3} = \frac{1}{3} + \frac{1}{3} + \frac{1}{3}$$

2. Complete the number bond and equation to match the tape diagram.

Tape Diagram	Number Bond and Equation
1	$1 = \dfrac{1}{4} + \dfrac{1}{4} + \dfrac{1}{4} + \dfrac{1}{4}$

The tape diagram represents 1.

The tape diagram is partitioned into 4 equal parts. Each part is 1 fourth, or $\frac{1}{4}$.

I can decompose 1 into unit fractions by writing $\frac{1}{4}$ in each part.

I can write 1 as the sum of the unit fractions.

$$1 = \frac{1}{4} + \frac{1}{4} + \frac{1}{4} + \frac{1}{4}$$

Copyright © Great Minds PBC

3. Draw and label a tape diagram to represent the equation.

Equation	Tape Diagram
$2 = \left(\frac{1}{2} + \frac{1}{2}\right) + \left(\frac{1}{2} + \frac{1}{2}\right)$	

I see in the equation that 2 is decomposed into 4 halves. The parentheses help me see that 2 halves make 1.

I draw a tape diagram to represent 2 and partition it into 2 equal parts.

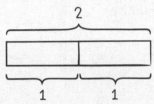

I can partition each 1 into 2 equal parts to show halves. Now, each part represents $\frac{1}{2}$.

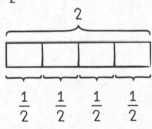

Copyright © Great Minds PBC

REMEMBER

4. Luke has 6 feet of ribbon. He cuts it into 4 equal pieces. How many inches long is each piece?

Each piece of ribbon is 18 inches long.

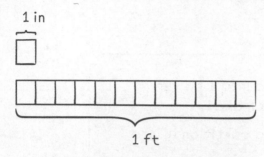

I know 1 foot = 12 inches.

1 in

1 ft

I can multiply 6 feet by 12 to find the total number of inches.

$$6 \times 12 = 72$$

Now I know the length of the ribbon in inches. It is 72 inches.

I can divide 72 by 4 to find the length of each piece of ribbon.

$$72 \div 4 = 18$$

A foot is 12 times as long as 1 inch.

Each interval of 1 foot on the number line represents 12 inches. I start at 60 inches and skip-count by twelves.

I see that there are 96 inches in 8 feet.

I am trying to find how many inches are in 8 feet 4 inches. I can add 4 more inches.

$$96 + 4 = 100$$

There are 100 inches in 8 feet 4 inches.

5. Fill in the blanks to complete the number line and the statement.

How many inches are in 8 feet 4 inches?

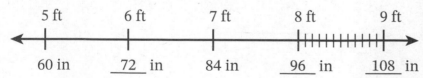

| 5 ft | 6 ft | 7 ft | 8 ft | 9 ft |
| 60 in | __72__ in | 84 in | __96__ in | __108__ in |

There are ___100___ inches in 8 feet 4 inches.

Copyright © Great Minds PBC

Name _____   Date _____

1. Complete the equations to match the partitioned circle. The circle represents 1.

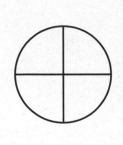

4 fourths = _____ fourth + _____ fourth + _____ fourth + _____ fourth

$$\frac{4}{4} = \frac{\blacksquare}{4} + \frac{\blacksquare}{4} + \frac{\blacksquare}{4} + \frac{\blacksquare}{4}$$

2. Complete the number bond and equation to match the tape diagram.

Tape Diagram	Number Bond and Equation

$$2 = (\underline{\quad} + \underline{\quad} + \underline{\quad}) + (\underline{\quad} + \underline{\quad} + \underline{\quad})$$

3. Draw and label a tape diagram to represent the equation.

Equation	Tape Diagram
$1 = \frac{1}{6} + \frac{1}{6} + \frac{1}{6} + \frac{1}{6} + \frac{1}{6} + \frac{1}{6}$	

Copyright © Great Minds PBC

REMEMBER

4. Ivan has 4 feet of rope. He cuts it into 6 equal pieces. How many inches long is each piece?

5. Fill in the blanks to complete the number line and the statement.

   How many inches are in 5 feet 7 inches?

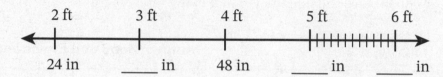

   There are _____ inches in 5 feet 7 inches.

Copyright © Great Minds PBC

Name

Date

1. Draw a tape diagram to represent the fraction. Then write an equation to express the fraction as a sum of unit fractions.

	Tape Diagram	**Equation**
$\frac{3}{5}$	1 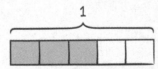	$\frac{3}{5} = \frac{1}{5} + \frac{1}{5} + \frac{1}{5}$

I draw and partition a tape diagram into 5 equal parts to show fifths.

Each part represents $\frac{1}{5}$. I can shade 3 parts to represent $\frac{3}{5}$.

1

I write an equation to show $\frac{3}{5}$ as a sum of unit fractions.

$$\frac{3}{5} = \frac{1}{5} + \frac{1}{5} + \frac{1}{5}$$

2. Draw a tape diagram to represent the fraction. Locate the fraction on the number line.

   Then write an equation to express the fraction as a sum of unit fractions.

Tape Diagram and Number Line	Equation
$\frac{8}{5}$ 	$\frac{8}{5} = \left(\frac{1}{5} + \frac{1}{5} + \frac{1}{5} + \frac{1}{5} + \frac{1}{5}\right) + \frac{1}{5} + \frac{1}{5} + \frac{1}{5}$

I can draw and partition a tape diagram to show 2 equal parts. Then I can partition each part into fifths.

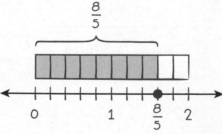

I can shade 8 of the parts to show the fraction $\frac{8}{5}$.

Below the tape diagram, I can draw a number line from 0 to 2. I want to end my number line at 2 because $\frac{8}{5}$ is greater than 1. I partition and label the number line to match the tape diagram.

I express the fraction $\frac{8}{5}$ as a sum of unit fractions. I use parentheses to show that $\frac{5}{5}$ is equal to 1. That helps me see that $\frac{8}{5}$ is greater than 1.

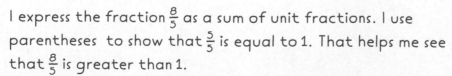

$$\frac{8}{5} = \left(\frac{1}{5} + \frac{1}{5} + \frac{1}{5} + \frac{1}{5} + \frac{1}{5}\right) + \frac{1}{5} + \frac{1}{5} + \frac{1}{5}$$

Copyright © Great Minds PBC

REMEMBER

3. Use the associative property to find factors of 42.

$$42 = 7 \times (\underline{}6\underline{})$$
$$= 7 \times (2 \times \underline{}3\underline{})$$
$$= (\underline{}7\underline{} \times 2) \times 3$$
$$= \underline{}14\underline{} \times 3$$
$$= \underline{}42\underline{}$$

Some factors of 42 are 1, 2, 3, 6, 7, 14, 42.

I know that 42 = 7 × 6, so 7 and 6 are factors.

I can use the associative property to help me find more factors.

I start with 42 = 7 × 6. I decompose 6 into 2 × 3 and change the grouping of the factors.

I multiply 2 by 7 to find that 42 = 14 × 3, so 3 and 14 are factors.

I also list 2 as a factor because 42 = 7 × 2 × 3.

I know that two factors of 42 are 1 and the number itself. So 1 and 42 are factors.

4. Think about the multiples of 3.

   a. Write the first 10 multiples of 3. Start with 3.

   3, 6, 9, 12, 15, 18, 21, 24, 27, 30

   b. What is the eighth multiple of 3? 24

   c. Is 28 a multiple of 3?

   No.

I can skip-count by threes to help me list the multiples.

I see that the eighth multiple of 3 is 24. I can also multiply 8 and 3 to check that the eighth multiple is 24.

I know that 28 is not a multiple of 3 because I don't say 28 when I count by threes.

Copyright © Great Minds PBC

Name _____ Date _____

1. Draw a tape diagram to represent the fraction. Then write an equation to express the fraction as a sum of unit fractions.

	Tape Diagram	Equation
$\frac{6}{8}$		

Draw a tape diagram to represent the fraction. Locate the fraction on the number line.

Then write an equation to express the fraction as a sum of unit fractions.

	Tape Diagram and Number Line	Equation
2. $\frac{5}{3}$		
3. $\frac{7}{4}$		

REMEMBER

4. Use the associative property to find factors of 60.

$$60 = 6 \times \underline{\hspace{1cm}}$$
$$= (3 \times \underline{\hspace{1cm}}) \times (2 \times \underline{\hspace{1cm}})$$
$$= (2 \times 2) \times (3 \times \underline{\hspace{1cm}})$$
$$= 4 \times \underline{\hspace{1cm}}$$
$$= \underline{\hspace{1cm}}$$

Some factors of 60 are _____.

5. Think about the multiples of 4.

a. Write the first 10 multiples of 4. Start with 4.

_____, _____, _____, _____, _____, _____, _____, _____, _____, _____

b. What is the seventh multiple of 4? _____

c. Is 32 a multiple of 4?

Copyright © Great Minds PBC

3

Name _____   Date _____

1. The following fraction is decomposed two different ways. Show how the fraction is decomposed
    by using a tape diagram, a number bond, and an equation.

	Tape Diagram	Number Bond	Equation
a. $\frac{6}{5}$	$\frac{6}{5}$ [tape divided, braces $\frac{4}{5}$ and $\frac{2}{5}$]	$\frac{6}{5}$ → $\frac{4}{5}$, $\frac{2}{5}$	$\frac{6}{5} = \underline{\frac{4}{5}} + \underline{\frac{2}{5}}$
b. $\frac{6}{5}$	$\frac{6}{5}$ [tape divided, braces $\frac{1}{5}$ and $\frac{5}{5}$]	$\frac{6}{5}$ → $\frac{1}{5}$, $\frac{5}{5}$	$\frac{6}{5} = \underline{\frac{1}{5}} + \underline{\frac{5}{5}}$

The tape diagram represents $\frac{6}{5}$. Each equal part represents $\frac{1}{5}$.

In part (a), I see the tape diagram is decomposed into two parts, $\frac{4}{5}$ and $\frac{2}{5}$. I show the two parts, $\frac{4}{5}$ and $\frac{2}{5}$, in the number bond and the equation.

In part (b), I see the tape diagram is decomposed differently into two parts, $\frac{1}{5}$ and $\frac{5}{5}$. I show the two parts, $\frac{1}{5}$ and $\frac{5}{5}$, in the number bond and the equation.

Copyright © Great Minds PBC

2. The following fraction is decomposed two different ways. Show how the fraction is decomposed by completing each equation.

a. $\frac{8}{6}$

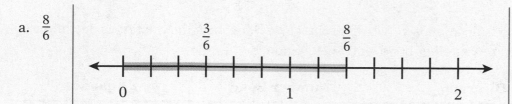

$$\frac{8}{6} = \underline{\quad \frac{3}{6} \quad} + \underline{\quad \frac{5}{6} \quad}$$

b. $\frac{8}{6}$

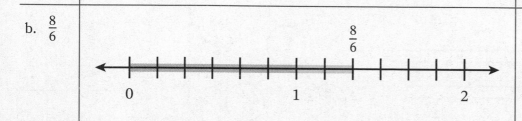

$$\frac{8}{6} = \underline{\quad \frac{6}{6} \quad} + \underline{\quad \frac{2}{6} \quad}$$

Each whole number interval on the number line is partitioned into sixths. Each interval represents $\frac{1}{6}$.

In part (a), 3 sixths are shaded pink, and 5 sixths are shaded yellow. The number line shows that $\frac{8}{6}$ is decomposed into $\frac{3}{6}$ and $\frac{5}{6}$.

In part (b), the number line shows another way to decompose $\frac{8}{6}$.

6 sixths are shaded pink and 2 sixths are shaded yellow. This number line shows that $\frac{8}{6}$ is decomposed into $\frac{6}{6}$ and $\frac{2}{6}$.

Copyright © Great Minds PBC

REMEMBER

3. The width of a rectangular floor is 8 feet. The area is 72 square feet. What is the length?

    The length of the floor is 9 feet.

> I know the width and area of the floor. I need to find the length of the floor.
>
> The formula for finding the area of a rectangle is $A = l \times w$.
>
> $$72 = l \times 8$$
>
> To find the length, I can think about what number I multiply by 8 to get 72.
>
> $$9 \times 8 = 72$$
>
> The length is 9 feet.

4. Multiply. Show your method.

$$3 \times 2{,}045 = \underline{6{,}135}$$

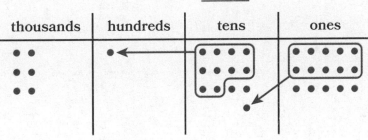

$$6{,}000 + 100 + 30 + 5 = 6{,}135$$

> I can use a place value chart. I know that 3 × 2,045 is 3 groups of 2,045. I draw 3 equal groups of 2,045.

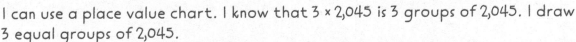

> I can rename 10 ones as 1 ten and rename 10 tens as 1 hundred.
>
> I count the units in each column. There are 6 thousands 1 hundred 3 tens 5 ones.
>
> I add the partial products to find the final product of 3 × 2,045.

Copyright © Great Minds PBC

Name _____ Date _____

3

1. The following fraction is decomposed two different ways. Show how the fraction is decomposed by using a tape diagram, a number bond, and an equation.

	Tape Diagram	Number Bond	Equation
a. $\frac{4}{3}$	$\frac{4}{3}$	$\frac{4}{3}$	$\frac{4}{3} =$ _____ $+$ _____
b. $\frac{4}{3}$	$\frac{4}{3}$	$\frac{4}{3}$	$\frac{4}{3} =$ _____ $+$ _____

2. The following fraction is decomposed two different ways. Show how the fraction is decomposed by completing each equation.

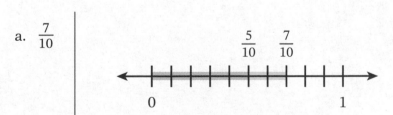

a. $\frac{7}{10}$ $\frac{5}{10}$ $\frac{7}{10}$

0   1

$\frac{7}{10} =$ _____ $+$ _____

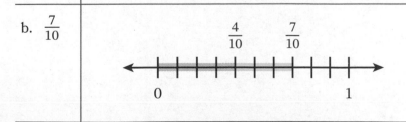

b. $\frac{7}{10}$ $\frac{4}{10}$ $\frac{7}{10}$

0   1

$\frac{7}{10} =$ _____ $+$ _____

REMEMBER

3. The width of a rectangular patio is 6 feet. The area is 48 square feet. What is the length?

4. Multiply. Show your method.

$$5,142 \times 3 = \underline{\hspace{2cm}}$$

thousands	hundreds	tens	ones

Copyright © Great Minds PBC

Name _____ Date _____

1. Draw a number line to represent the fractional amount. Write an equation to show one way to decompose the fraction.

	Number Line	Equation
$\frac{5}{8}$ hours 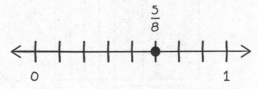	$\frac{5}{8}$ number line from 0 to 1	$\frac{5}{8} = \frac{2}{8} + \frac{3}{8}$

The tape diagram represents 1. The tape diagram is partitioned into 8 equal parts.

5 parts, or 5 eighths, are shaded. So the tape diagram shows $\frac{5}{8}$ hours.

I can draw a number line from 0 to 1 and partition it into eighths.

I plot a point at the tick mark representing $\frac{5}{8}$.

$\frac{5}{8}$

I can decompose the fraction $\frac{5}{8}$ into two parts and write an equation.

$$\frac{5}{8} = \frac{2}{8} + \frac{3}{8}$$

Copyright © Great Minds PBC

2. Draw an area model to represent the fractional amount. Write an equation to show one way to decompose the fraction.

	Area Model	Equation
$\frac{6}{10}$ hours 	(area model partitioned into 10 equal parts with 6 shaded, labeled 1 on top and $\frac{6}{10}$ on bottom)	$\frac{6}{10} = \frac{1}{10} + \frac{5}{10}$

I partition an area model into 10 equal parts. So each part represents $\frac{1}{10}$.

I can shade 6 parts to show the fraction $\frac{6}{10}$.

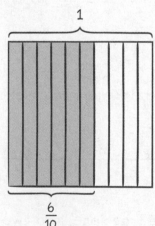

I can decompose the fraction $\frac{6}{10}$ into two parts and write an equation.

$$\frac{6}{10} = \frac{1}{10} + \frac{5}{10}$$

Copyright © Great Minds PBC

REMEMBER

3. Complete the table.

Yards	Feet
15	45
17	51
19	57
20	60

4. A playground is 17 yards long. Write the length of the playground in feet.

   51 feet

I know 1 yard is 3 times as long as 1 foot.

To convert yards to feet, I can multiply the number of yards by 3.

$$15 \times 3 = 45$$

$$17 \times 3 = 51$$

$$19 \times 3 = 57$$

$$20 \times 3 = 60$$

The playground is 51 feet long.

Name _____ Date _____

1. Draw a number line to represent the fractional amount. Write an equation to show one way to decompose the fraction.

	Number Line	Equation
$\frac{4}{5}$ miles 1 		

2. Draw an area model to represent the fractional amount. Write an equation to show one way to decompose the fraction.

	Area Model	Equation
$\frac{5}{6}$ parts of a pie 		

3. Draw two different models to represent $\frac{2}{3}$ hours.

REMEMBER

4. Complete the table.

Yards	Feet
10	30
15	____
20	____
21	63

5. A fence is 20 yards long. Write the length of the fence in feet.

    _____ feet

Copyright © Great Minds PBC

Name _____   Date _____

5

1. Complete the number bond and equation. Rename the fractions that are equivalent to whole numbers.

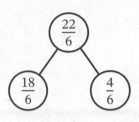

$$\frac{22}{6} = \frac{18}{6} + \frac{4}{6}$$

$$\frac{22}{6} = \underline{\quad 3 \quad} + \frac{4}{6}$$

$$\frac{22}{6} = 3\,\frac{4}{6}$$

I need to write the fraction $\frac{22}{6}$ as a mixed number. A **mixed number** is a whole number and a fraction less than 1.

I know $1 = \frac{6}{6}$. So that means $3 = \frac{18}{6}$.

I can decompose $\frac{22}{6}$ into two parts. I know one part is $\frac{18}{6}$, so the other part is $\frac{4}{6}$.

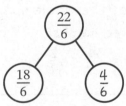

I can rename $\frac{18}{6}$ as 3.

$$\frac{22}{6} = 3 + \frac{4}{6}$$

I can write the sum of the whole number, 3, and the fraction, $\frac{4}{6}$, as a mixed number.

$$\frac{22}{6} = 3\frac{4}{6}$$

2. Rename the fraction greater than 1 as a mixed number. Use a model to show the decomposition.

$$\frac{30}{7} = 4\frac{2}{7}$$

$$\frac{30}{7} = \frac{28}{7} + \frac{2}{7}$$

$$\frac{30}{7} = 4 + \frac{2}{7}$$

$$\frac{30}{7} = 4\frac{2}{7}$$

I need to write the fraction $\frac{30}{7}$ as a mixed number.

I know $1 = \frac{7}{7}$. So that means $4 = \frac{28}{7}$.

I can decompose $\frac{30}{7}$ into two parts. I know one part is $\frac{28}{7}$, so the other part is $\frac{2}{7}$. I can draw a number bond to show the decomposition.

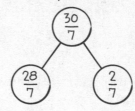

I can write the sum of the whole number, 4, and the fraction, $\frac{2}{7}$, as a mixed number.

$$\frac{30}{7} = 4\frac{2}{7}$$

Copyright © Great Minds PBC

REMEMBER

3. A company wants to donate books to a local elementary school. They want to be sure every student gets a new book. There are 128 students in the lower grades. There are 144 students in the middle grades. There are 196 students in the upper grades.

   a. The company estimates the total number of students by rounding each number to the nearest hundred. What is their estimate?

   $$100 + 100 + 200 = 400$$

   The total number of students is 400.

   b. The company thinks they estimated correctly. What is the actual total number of students?

   $$128 + 144 + 196 = 468$$

   The actual total number of students is 468.

   c. Will the company's estimate provide enough books for every student?

   No.

   d. To make sure the company provides enough books, what strategy can they use to estimate?

   The company can round each number of students to the next ten to make sure they send enough books to the school. Rounding to the next ten means that there will be will extra books. If they just round to the nearest ten, their estimate might be lower than the actual number of books that is needed.

When I round to the nearest hundred, my estimate is not useful because it would mean that 68 students would not get books.

I can get a closer estimate if I round to the nearest ten, but I can't be sure that the estimate will be higher than the actual total. So, I round to the next ten which means my estimate will be greater than the actual total.

Name _____   Date _____

Complete each number bond and equation. Rename the fractions that are equivalent to whole numbers.

1.
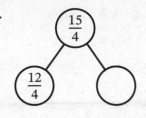

$$\frac{15}{4} = \frac{12}{4} + \frac{\boxed{}}{\boxed{}}$$

$$\frac{15}{4} = \underline{\hspace{1cm}} + \frac{\boxed{}}{\boxed{}}$$

$$\frac{15}{4} = \boxed{} \frac{\boxed{}}{\boxed{}}$$

2.
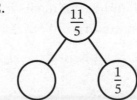

$$\frac{11}{5} = \frac{\boxed{}}{5} + \frac{1}{5}$$

$$\frac{11}{5} = \underline{\hspace{1cm}} + \frac{\boxed{}}{\boxed{}}$$

$$\frac{11}{5} = \boxed{} \frac{\boxed{}}{\boxed{}}$$

Rename each fraction greater than 1 as a mixed number. Use a model to show the decomposition.

3. $\frac{35}{8} =$

4. $\frac{67}{12} =$

REMEMBER

5. Shen is ordering meals for a picnic for 3 local schools. He wants to estimate the number of meals to order. Shen needs meals for 124 students from East Elementary School, 178 students from North Elementary School, and 206 students from Central Elementary School.

   a. Shen estimates the total number of meals for all 3 schools by rounding each number of students to the nearest hundred. What is his estimate?

   b. Shen thinks he estimated correctly. What is the actual total number of students?

   c. Will Shen's estimate provide enough meals?

   d. To make sure Shen orders enough meals, what strategy can he use to estimate?

Copyright © Great Minds PBC

6

Name _____ Date _____

Rename each mixed number as a fraction greater than 1. Show the decomposition on the number line and write the equation.

1. $2\frac{1}{6}$

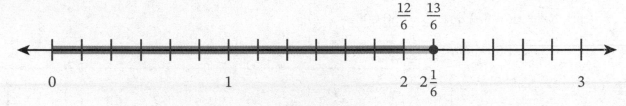

$$2\frac{1}{6} = \frac{12}{6} + \frac{1}{6}$$
$$2\frac{1}{6} = \frac{13}{6}$$

I know the mixed number $2\frac{1}{6}$ is between 2 and 3. I can represent 0 to 3 on the number line. I partition the number line into three equal parts and label the whole numbers 0, 1, 2, and 3.

The fractional unit is sixths. So I partition each whole number interval into six equal parts.

I can decompose $2\frac{1}{6}$ into two parts. The first part is the whole number, 2. I know $2 = \frac{12}{6}$. The second part is the fraction, $\frac{1}{6}$. I shade from 0 to 2 in one color. Then I use a second color to shade from 2 to the next tick mark. Both colors together show $2\frac{1}{6}$.

I write the equation $2\frac{1}{6} = \frac{12}{6} + \frac{1}{6}$ to show the decomposition of $2\frac{1}{6}$. I can rename the mixed number as a fraction greater than 1. $2\frac{1}{6}$ is $\frac{1}{6}$ more than $\frac{12}{6}$, so $2\frac{1}{6} = \frac{13}{6}$.

Copyright © Great Minds PBC

REMEMBER

2. Round to the given place.

648,061

a. Nearest hundred thousand 600,000

b. Nearest ten thousand 650,000

c. Nearest thousand 648,000

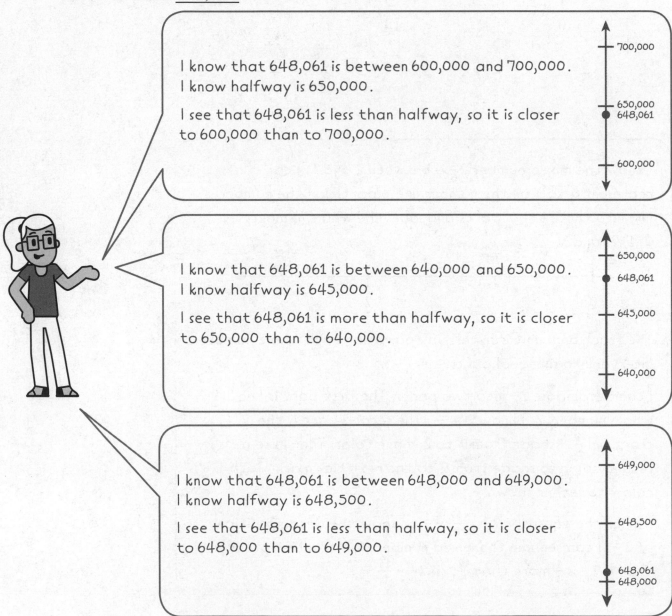

I know that 648,061 is between 600,000 and 700,000. I know halfway is 650,000.

I see that 648,061 is less than halfway, so it is closer to 600,000 than to 700,000.

I know that 648,061 is between 640,000 and 650,000. I know halfway is 645,000.

I see that 648,061 is more than halfway, so it is closer to 650,000 than to 640,000.

I know that 648,061 is between 648,000 and 649,000. I know halfway is 648,500.

I see that 648,061 is less than halfway, so it is closer to 648,000 than to 649,000.

Copyright © Great Minds PBC

Name _____ Date _____

1. Rename the mixed number as a fraction greater than 1. Show the decomposition on the number line and write the equation.

   $2\frac{3}{4}$

Rename each mixed number as a fraction greater than 1. Use a model to show the decomposition.

2. $1\frac{2}{3}$

3. $3\frac{5}{6}$

REMEMBER

4. Round to the given place.

$$62,538$$

a. Nearest hundred thousand _____

b. Nearest ten thousand _____

c. Nearest thousand _____

Copyright © Great Minds PBC

FAMILY MATH
Equivalent Fractions

Dear Family,

Your student is learning to write equivalent fractions by renaming the units, or denominator, of the fraction. They use models such as tape diagrams, area models, and number lines along with multiplication and division to make equivalent fractions. They also learn to write equivalent fractions for fractions greater than 1 and for mixed numbers. Finding equivalent fractions prepares your student for comparing, adding, and subtracting fractions in future lessons.

Key Terms

denominator

numerator

$$\frac{1}{2} \begin{array}{l} \longleftarrow \text{ Numerator} \\ \longleftarrow \text{ Denominator} \end{array}$$

$$\frac{1}{4} = \left(\frac{1}{12} + \frac{1}{12} + \frac{1}{12} \right)$$

$$\frac{1}{4} = \frac{3}{12}$$

Students learn that the denominator of a fraction indicates the total number of units in the whole and the numerator indicates the number of selected parts.

Students break apart fractions into smaller unit fractions. They can add the smaller unit fractions together to make equivalent fractions.

$$\frac{1}{4} = \frac{3 \times 1}{3 \times 4} = \frac{3}{12}$$

$$\frac{2}{3} = \frac{2 \times 2}{2 \times 3} = \frac{4}{6}$$

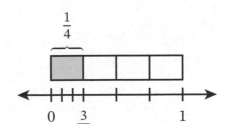

$$\frac{9}{12} = \frac{9 \div 3}{12 \div 3} = \frac{3}{4}$$

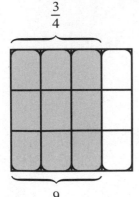

Area models and number lines help students see how the number of parts and the size of the parts change when multiplication or division is used to create an equivalent fraction.

Copyright © Great Minds PBC

At-Home Activities

Naming Numerators and Denominators

Help your student practice by using the terms *numerator* and *denominator* to describe the parts of fractions. Look for fractions in your daily activities, such as when measuring ingredients to follow recipes. Ask your student to identify the numerator, which is the number of selected parts. Ask your student to identify the denominator, which is the total number of units in the whole. Then discuss what the numerator and the denominator each represent.

What is Half?

With your student, practice naming 1 half of something in different situations. Consider using the following examples.

- There are 4 quarters in a dollar. What is half of this amount? (2 quarters) What fraction of 4 quarters is 2 quarters? $\left(\frac{2}{4} \text{ or } \frac{1}{2}\right)$

- There are 8 ounces in a measuring cup. What is half of this amount? (4 ounces) What fraction of 8 ounces is 4 ounces? $\left(\frac{4}{8} \text{ or } \frac{1}{2}\right)$

- There are 12 months in a year. How many months are in half of a year? (6 months) What fraction of 12 months is 6 months? $\left(\frac{6}{12} \text{ or } \frac{1}{2}\right)$

- There are 60 minutes in an hour. How many minutes are in half of an hour? (30 minutes) What fraction of 60 minutes is 30 minutes? $\left(\frac{30}{60} \text{ or } \frac{1}{2}\right)$

     Copyright © Great Minds PBC

7

Name _____ Date _____

1. Rename the fraction by decomposing it into smaller unit fractions.

   Label the number line to show the decomposition. Then complete the equations.

   Number Line and Equation

   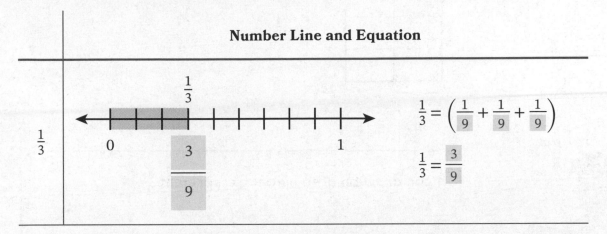

   $\frac{1}{3}$

   $\frac{1}{3} = \left(\dfrac{1}{9} + \dfrac{1}{9} + \dfrac{1}{9} \right)$

   $\frac{1}{3} = \dfrac{3}{9}$

   I see $\frac{1}{3}$ labeled on the number line.

   $\frac{1}{3}$ is decomposed into three equal parts to make ninths.

   The parentheses show how many of the smaller unit fractions equal 1 of the larger unit fraction.

   $\frac{1}{3}$ is equivalent to $\frac{3}{9}$.

Copyright © Great Minds PBC

2. Show that the fractions are equivalent by drawing a model. Then write an equation to express the equivalence as a sum.

$\frac{1}{2}$ and $\frac{3}{6}$

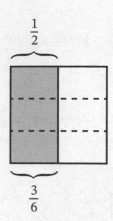

$$\frac{1}{2} = \frac{1}{6} + \frac{1}{6} + \frac{1}{6}$$

I can draw an area model to represent $\frac{1}{2}$.

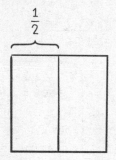

I can draw 2 horizontal lines to partition each half into 3 equal parts. There are now 6 equal parts. 3 of the parts are shaded.

I can write this as the fraction $\frac{3}{6}$.

So $\frac{1}{2} = \frac{3}{6}$.

Copyright © Great Minds PBC

REMEMBER

3. Record the factor pair for the given number as a multiplication expression. List the factors in order from least to greatest. Then circle prime or composite for the number.

Number	Multiplication Expression	Factors	Prime or Composite
13	1×13	1, 13	(Prime) Composite

A whole number greater than 1 is a prime number if its only factors are 1 and itself.

A whole number greater than 1 is a composite number if it has more than two factors.

I can make rectangular arrays of 13 circles to find the factor pairs.

One array has 1 row of 13 circles which shows that $1 \times 13 = 13$.

○ ○ ○ ○ ○ ○ ○ ○ ○ ○ ○ ○ ○

There are no other rectangular arrays that can be made with 13 circles.

The only factor pair for 13 is 1×13.

Because the only factors are 1 and 13, the number is prime.

Copyright © Great Minds PBC

Name _____   Date _____

1. Rename the fraction by decomposing it into smaller unit fractions.

   Label the number line to show the decomposition. Then complete the equations.

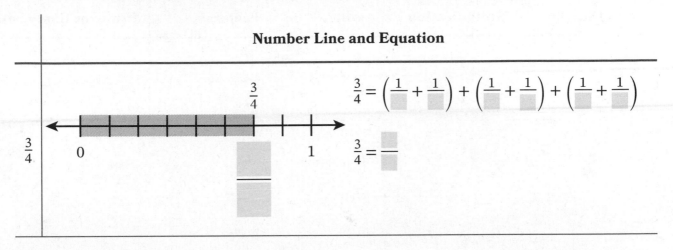

Number Line and Equation

$$\frac{3}{4} = \left(\frac{1}{\Box} + \frac{1}{\Box}\right) + \left(\frac{1}{\Box} + \frac{1}{\Box}\right) + \left(\frac{1}{\Box} + \frac{1}{\Box}\right)$$

$$\frac{3}{4} = \frac{\Box}{\Box}$$

2. Show that the fractions are equivalent by drawing a model. Then write an equation to express the equivalence as a sum.

   $\frac{2}{4}$ and $\frac{6}{12}$

REMEMBER

3. Record the factor pair for the given number as a multiplication expression. List the factors in order from least to greatest. Then circle prime or composite for the number.

Number	Multiplication Expression	Factors	Prime or Composite
17			Prime Composite

Copyright © Great Minds PBC

Name _____ Date _____

1. Partition the area model to show smaller fractional units. Then label the equivalent fraction. Express the equivalence by using multiplication.

   Decompose into tenths.

   $\frac{1}{2}$

   $\frac{1}{2} = \frac{5 \times 1}{5 \times 2} = \frac{5}{10}$

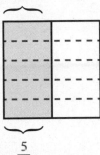

   $\frac{5}{10}$

   The area model is partitioned into 2 equal parts. 1 of the parts is shaded. This represents the fraction $\frac{1}{2}$.

   I can partition each half with horizontal lines to make 5 times as many parts. There are now 10 equal parts. Each part is $\frac{1}{10}$.

   I multiply the numerator, 1, and the denominator, 2, by 5 to make an equivalent fraction, $\frac{5}{10}$.

   I draw an area model to represent $\frac{1}{5}$. I shade one of the parts. This represents the fraction $\frac{1}{5}$.

   $\frac{1}{5}$

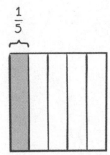

2. Show that the fractions are equivalent by drawing an area model. Then express the equivalence by using multiplication.

   $\frac{1}{5}$ and $\frac{2}{10}$

   $\frac{1}{5}$

   $\frac{1}{5} = \frac{2 \times 1}{2 \times 5} = \frac{2}{10}$

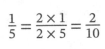

   $\frac{2}{10}$

   I can partition each fifth horizontally to make 2 times as many parts. There are now 10 equal parts.

   2 of the 10 parts are shaded. I can write this as the fraction $\frac{2}{10}$.

   I multiply the numerator, 1, and the denominator, 5, by 2 to make an equivalent fraction, $\frac{2}{10}$.

Copyright © Great Minds PBC

REMEMBER

3. On Monday, Ray reads 4 pages in a book. Each day, he reads 4 more pages than the day before.

   a. Complete the table.

Day	Monday	Tuesday	Wednesday	Thursday	Friday
Number of Pages	4	8	12	16	20

   b. What patterns do you notice in the number of pages?

   Sample: Every other number in the pattern is a multiple of 8. I also notice that each number is even.

I can skip-count by fours to complete the table.

I look for another pattern.

I know that 4 is a factor of 8, and 2 × 4 = 8. So every other multiple of 4 will be a multiple of 8.

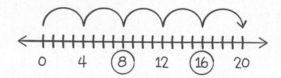

Copyright © Great Minds PBC

Name

Date

1. Partition the area model to show smaller fractional units. Then label the equivalent fraction. Express the equivalence by using multiplication.

Decompose into twelfths.

$\frac{1}{4}$

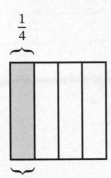

$$\frac{1}{4} = \frac{\blacksquare \times 1}{\blacksquare \times 4} = \frac{\blacksquare}{\blacksquare}$$

2. Show that the fractions are equivalent by drawing an area model. Then express the equivalence by using multiplication.

$\frac{1}{2}$ and $\frac{3}{6}$

Copyright © Great Minds PBC

REMEMBER

3. On Monday, Jayla jumps rope for 3 minutes. Each day, she jumps rope for 2 more minutes than the day before.

   a. Complete the table.

Day	Monday	Tuesday	Wednesday	Thursday	Friday
Number of Minutes	3				

   b. What pattern do you notice in the number of minutes?

Copyright © Great Minds PBC

Name _____

Date _____

1. The square represents 1.

   Label the area model to show the equivalent fraction.

   Express the equivalence by using multiplication.

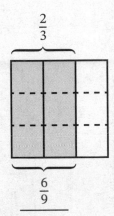

$$\frac{2}{3} = \frac{3 \times 2}{3 \times 3} = \frac{6}{9}$$

The area model was partitioned into 3 equal parts, or thirds. Then, the area model was partitioned into 3 times as many parts. The partitions created 3 times as many shaded parts. I see $\frac{6}{9}$ is equivalent to $\frac{2}{3}$.

Copyright © Great Minds PBC

2. Partition the area model to show smaller units. Then label the equivalent fraction.

Express the equivalence by using multiplication.

Decompose into twelfths.

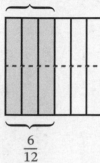

$$\frac{3}{6} = \frac{2 \times 3}{2 \times 6} = \frac{6}{12}$$

The area model is partitioned into 6 equal parts. 3 of the parts are shaded. This represents the fraction $\frac{3}{6}$.

I can partition each sixth with a horizontal line to make 2 times as many parts. There are now 12 equal parts. Each part is $\frac{1}{12}$.

Now 6 units of $\frac{1}{12}$ are shaded. I can write this as the fraction $\frac{6}{12}$.

I multiply the numerator, 3, and the denominator, 6, by 2 to make an equivalent fraction, $\frac{6}{12}$.

Copyright © Great Minds PBC

REMEMBER

3. Complete the conversion table.

Hours	Minutes
1	60
4	240
8	480
11	660
16	960

I know 1 hour is equal to 60 minutes.

To convert hours to minutes, I can multiply the number of hours by 60. I can use unit form to help me multiply.

$$4 \times 6 \text{ tens} = 24 \text{ tens}$$
$$4 \times 60 = 240$$

$$8 \times 6 \text{ tens} = 48 \text{ tens}$$
$$8 \times 60 = 480$$

$$11 \times 6 \text{ tens} = 66 \text{ tens}$$
$$11 \times 60 = 660$$

$$16 \times 6 \text{ tens} = 96 \text{ tens}$$
$$16 \times 60 = 960$$

Name   Date

1. The square represents 1.

    Label the area model to show the equivalent fraction.

    Express the equivalence by using multiplication.

$$\frac{4}{5}$$

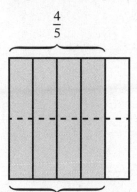

$$\frac{4}{5} = \frac{2 \times 4}{2 \times 5} = \frac{\blacksquare}{\blacksquare}$$

_____

2. Partition the area model to show smaller units. Then label the equivalent fraction.

    Express the equivalence by using multiplication.

    Decompose into eighths.

$$\frac{3}{4}$$

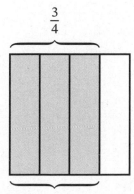

$$\frac{3}{4} =$$

_____

REMEMBER

3. Complete the conversion table.

Hours	Minutes
1	
2	
6	
18	
20	

Copyright © Great Minds PBC

Name _____   Date _____

1. The large square represents 1.

    Trace the area model to show how the larger unit is composed. Then label the equivalent fractions.

    Express the equivalence by using division.

$$\frac{1}{3}$$

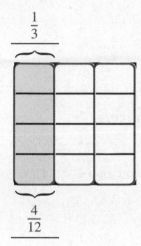

$$\frac{4}{12} = \frac{4 \div 4}{12 \div 4} = \frac{1}{3}$$

$$\frac{4}{12}$$

The area model shows 12 parts and 4 of the parts are shaded. The model represents $\frac{4}{12}$.

I can compose a larger unit by tracing each column. Each column represents $\frac{1}{3}$. There are 4 units of $\frac{1}{12}$ in $\frac{1}{3}$ of the area model.

I can also use division to write a fraction equivalent to $\frac{4}{12}$.

REMEMBER

2. Casey drives for 5 hours 15 minutes. She stops to eat lunch. Then Casey drives for another 25 minutes. How many total minutes does Casey drive?

$$5 \times 60 = 300$$

$$300 + 15 = 315$$

$$315 + 25 = 340$$

Casey drives for 340 minutes.

First, I convert the hours to minutes.

There are 60 minutes in 1 hour.

I can multiply the number of hours by 60 to find the total number of minutes.

$$5 \times 60 = 300$$

I add 300 and 15. Casey drives for 315 minutes before lunch.

After lunch, Casey drives for another 25 minutes. I add 315 and 25 to find how many minutes Casey drives in total.

Copyright © Great Minds PBC

Name _____   Date _____

/ 10

Each large square represents 1.

Trace the area model to show how the larger unit is composed. Then label the equivalent fractions.

Express each equivalence by using division.

1.

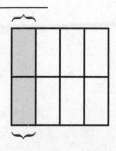

$$\frac{2}{8} = \frac{2 \div 2}{8 \div 2} = \frac{\ }{\ }$$

2.

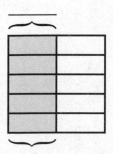

REMEMBER

3. Robin hikes for 3 hours 40 minutes. She stops for a break. Then Robin hikes for another 35 minutes. How many total minutes does Robin hike?

Copyright © Great Minds PBC

Name _____ Date _____

1. Partition, shade, and label the tape diagram to represent $\frac{5}{6}$. Then use the tape diagram and multiplication to partition and label the number line to represent the equivalent fraction.

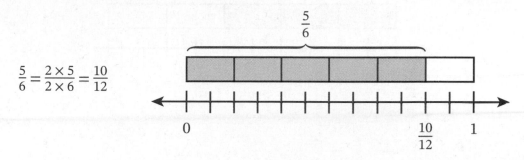

$$\frac{5}{6} = \frac{2 \times 5}{2 \times 6} = \frac{10}{12}$$

I partition the tape diagram into 6 equal parts. I shade 5 parts which represents the fraction $\frac{5}{6}$.

The equivalent fraction is $\frac{10}{12}$. Each sixth is decomposed into 2 equal parts to make twice as many units. I partition 1 whole into 12 equal parts on the number line.

5 units of $\frac{1}{6}$ on the tape diagram is equal to 10 units of $\frac{1}{12}$ on the number line. So $\frac{5}{6} = \frac{10}{12}$.

Copyright © Great Minds PBC

2. Partition and label the number line to represent $\frac{6}{10}$. Then use the number line and division to partition, shade, and label the tape diagram to represent the equivalent fraction.

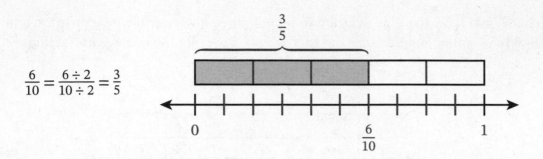

$$\frac{6}{10} = \frac{6 \div 2}{10 \div 2} = \frac{3}{5}$$

I partition 1 whole into 10 equal parts on the number line.

I label the seventh tick mark as $\frac{6}{10}$.

The equivalent fraction is $\frac{3}{5}$. I partition the tape diagram into 5 equal parts.

I compose 2 smaller units to make a larger unit. The number line and tape diagram help me see that $\frac{2}{10} = \frac{1}{5}$.

I shade 3 parts of the tape diagram which represents the fraction $\frac{3}{5}$.

        Copyright © Great Minds PBC

REMEMBER

3. How much time passes from 9:50 a.m. to 11:04 a.m.?

Start   **End**

____74____ minutes

I start at 9:50 and count the minutes to 10:00. That amount of time is 10 minutes.

I know from 10:00 to 11:00 is 60 minutes.

I also know from 11:00 to 11:04 is 4 minutes.

$$10 + 60 + 4 = 74$$

There are 74 minutes from 9:50 a.m. to 11:04 a.m.

Copyright © Great Minds PBC

Name _____ Date _____

1. Partition, shade, and label the tape diagram to represent $\frac{4}{5}$. Then use the tape diagram and multiplication to partition and label the number line to represent the equivalent fraction.

$$\frac{4}{5} = \frac{2 \times 4}{2 \times 5} = \frac{8}{10}$$

2. Partition and label the number line to represent $\frac{4}{12}$. Then use the number line and division to partition, shade, and label the tape diagram to represent the equivalent fraction.

$$\frac{4}{12} = \frac{4 \div 2}{12 \div 2} = \frac{2}{6}$$

3. Partition and label the number line to represent $\frac{4}{8}$. Then use the number line and division to partition, shade, and label the tape diagram to represent the equivalent fraction.

$$\frac{4}{8} = \frac{4 \div 2}{8 \div 2} = \frac{2}{4}$$

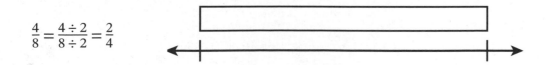

REMEMBER

4. How much time passes from 2:35 p.m. to 4:02 p.m.?

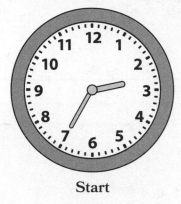
Start

End

_____ minutes

Copyright © Great Minds PBC

Name _____ Date _____

1. Label the number line to show the fraction greater than 1.

   Then use multiplication to express the equivalence.

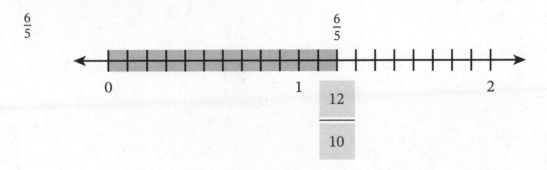

$\frac{6}{5}$

$\boxed{12}$

$\boxed{10}$

$$\frac{6}{5} = \frac{2 \times 6}{2 \times 5} = \frac{12}{10}$$

The fraction $\frac{6}{5}$ is greater than 1.

Each whole number interval is partitioned into tenths, which is 2 times as many units as fifths.

I can use multiplication to find a fraction written in tenths that is equivalent to $\frac{6}{5}$.

$$\frac{6}{5} = \frac{2 \times 6}{2 \times 5} = \frac{12}{10}$$

I write the fraction $\frac{12}{10}$ beneath the number line.
I know $\frac{6}{5}$ and $\frac{12}{10}$ are equivalent because they share the same location on the number line.

Copyright © Great Minds PBC

2. Label the number line to show the mixed number.

   Then use multiplication to express the equivalence.

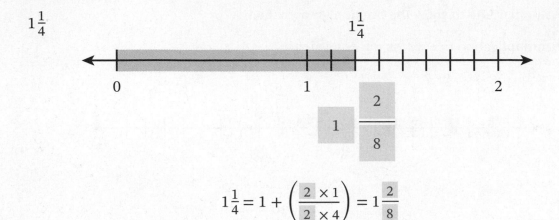

$$1\frac{1}{4} = 1 + \left(\frac{2 \times 1}{2 \times 4}\right) = 1\frac{2}{8}$$

The mixed number $1\frac{1}{4}$ is between 1 and 2.

The whole number interval 1 to 2 is partitioned into eighths, which is 2 times as many units as fourths.

I can use multiplication to find a fraction written in eighths that is equivalent to $\frac{1}{4}$.

$$\frac{1}{4} = \frac{2 \times 1}{2 \times 4} = \frac{2}{8}$$

I write the mixed number $1\frac{2}{8}$ beneath the number line. I know $1\frac{2}{8}$ is equivalent to $1\frac{1}{4}$ because they share the same location on the number line.

                    Copyright © Great Minds PBC

3. Complete the equation to show equivalent fractions.

   You may draw a number line to help you.

   $\dfrac{7}{4} = \dfrac{21}{12}$

   The fraction $\dfrac{7}{4}$ is equal to a number of twelfths.

   I can draw a number line from 0 to 2, partition each whole number interval into fourths, and label $\dfrac{7}{4}$. I can partition each fourth into 3 equal parts to make twelfths.

   Twelfths are 3 times the number of units as fourths.

   I multiply the numerator and the denominator by 3 to find a fraction written in twelfths that is equivalent to $\dfrac{7}{4}$.

   $$\dfrac{7}{4} = \dfrac{3 \times 7}{3 \times 4} = \dfrac{21}{12}$$

REMEMBER

4. David must wait 50 minutes before he can frost cupcakes. He waits from 11:45 a.m. to 12:07 p.m. How much longer must David wait?

   David must wait 28 more minutes.

   David waits from 11:45 a.m. to 12:07 p.m.

   I can find how long he waited by using the benchmark time 12:00 p.m.

   11:45 a.m. to 12:00 p.m. is 15 minutes.

   12:00 p.m. to 12:07 p.m. is 7 minutes.

   15 + 7 = 22, so David waited for 22 minutes.

   I know David must wait a total of 50 minutes.

   I can subtract 22 from 50 to find how much longer David must wait.

   David must wait 28 more minutes.

Name _____

Date _____

12

1. Label the number line to show fractions greater than 1.

   Then use multiplication to express the equivalence.

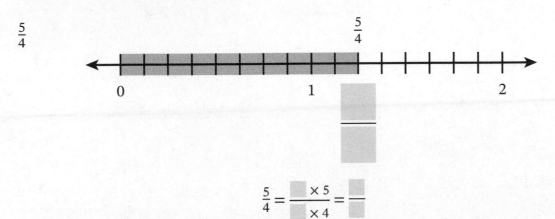

   $\dfrac{5}{4}$

   $\dfrac{5}{4}$

   $\dfrac{5}{4} = \dfrac{\boxed{} \times 5}{\boxed{} \times 4} = \dfrac{\boxed{}}{\boxed{}}$

2. Label the number line to show the mixed number.

   Then use multiplication to express the equivalence.

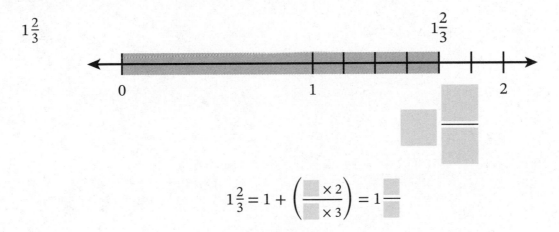

   $1\dfrac{2}{3}$

   $1\dfrac{2}{3}$

   $1\dfrac{2}{3} = 1 + \left(\dfrac{\boxed{} \times 2}{\boxed{} \times 3} \right) = 1\dfrac{\boxed{}}{\boxed{}}$

3. Complete the equation to show equivalent fractions. You may draw a number line to help you.

   $\dfrac{4}{3} = \dfrac{\boxed{}}{9}$

4. $3\dfrac{1}{2} = 3\dfrac{2}{\boxed{}}$

REMEMBER

5. Deepa needs to walk her dog for 30 minutes. She walks her dog from 3:53 p.m. to 4:14 p.m. How many more minutes does Deepa need to walk her dog?

Copyright © Great Minds PBC

FAMILY MATH
Comparing Fractions

Dear Family,

Your student is using models and equations to compare fractions. Your student estimates the locations of two or more fractions on a number line. They learn to compare unlike fractions by finding a common denominator and comparing the sizes of the numerators. Your student can also compare unlike fractions by finding a common numerator and comparing the sizes of the denominators. Comparing fractions will help your student when checking the reasonableness of their answers in future lessons.

Key Terms

common denominator

common numerator

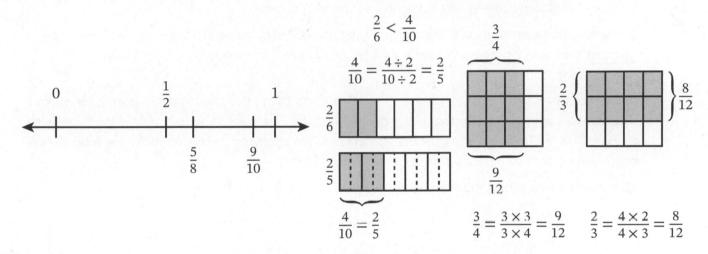

Students may use benchmark numbers, such as $0, \frac{1}{2}$, or 1, to help them determine where a fraction belongs on a number line. $\frac{5}{8}$ is less than $\frac{9}{10}$ because $\frac{5}{8}$ is closer to 0.

Sometimes students only have to rename one fraction to make a common numerator or denominator. When 4 tenths is renamed as 2 fifths, students can see that 2 fifths is larger because fifths are larger than sixths.

Sometimes students have to rename both fractions to make a common numerator or denominator. When both fractions have the same denominator, students can compare the numbers of units.

At-Home Activity

Who Has the Greater Fraction?

Practice comparing fractions by playing a card game.

- First, you will need to make your cards. On index cards or small pieces of paper, write 10 different fractions that have 2, 3, 4, 5, 6, 8, 10, 12, or 100 as a denominator. Then make a second set of 10 cards by using the same denominators. The second set can have the same cards as the first set, or they can be different.

- Mix up the cards in each set and give one set of fraction cards to your student. Place your set facedown in front of you and have your student do the same with their set of cards.

- Each person turns over their top card at the same time.

- Discuss which person has the fraction with the greater value. The person who has the greater fraction takes both cards and places them facedown on the bottom of their stack of cards.

- If the fractions are equal, then have each person turn over the next card in their stack. Repeat this step until the fractions are not equal. The person who has the fraction with the greater value then wins all the cards that are showing in this round. Place all these cards facedown on the bottom of the winner's deck.

- The game ends when one person has all the cards from each set.

                   Copyright © Great Minds PBC

Name _____ Date _____

1. Plot the fractions on the number line. Use >, =, or < to compare the fractions.

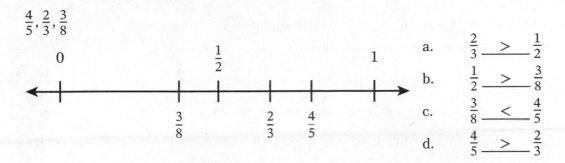

$\frac{4}{5}, \frac{2}{3}, \frac{3}{8}$

a. $\frac{2}{3}$ ___>___ $\frac{1}{2}$

b. $\frac{1}{2}$ ___>___ $\frac{3}{8}$

c. $\frac{3}{8}$ ___<___ $\frac{4}{5}$

d. $\frac{4}{5}$ ___>___ $\frac{2}{3}$

I know that $\frac{4}{5}$ and $\frac{2}{3}$ are between $\frac{1}{2}$ and 1.

Both fractions are 1 fractional unit away from 1.

$\frac{4}{5}$ is closer to 1 than $\frac{2}{3}$ because a fifth is a smaller unit than a third.

I know $\frac{4}{8}$ is equivalent to $\frac{1}{2}$.

$\frac{3}{8}$ is 1 fractional unit away from $\frac{4}{8}$. Because $\frac{3}{8}$ is less than $\frac{4}{8}$, I plot $\frac{3}{8}$ between 0 and $\frac{1}{2}$, so it is closer to $\frac{1}{2}$.

I use the number line to help me compare the fractions.

I see that $\frac{2}{3}$ is closer to 1 than $\frac{1}{2}$, so $\frac{2}{3} > \frac{1}{2}$.

I see that $\frac{1}{2}$ is closer to 1 than $\frac{3}{8}$, so $\frac{1}{2} > \frac{3}{8}$.

I see that $\frac{3}{8}$ is less than $\frac{1}{2}$ and $\frac{4}{5}$ is greater than $\frac{1}{2}$, so $\frac{3}{8} < \frac{4}{5}$.

I see that $\frac{4}{5}$ is closer to 1 than $\frac{2}{3}$, so $\frac{4}{5} > \frac{2}{3}$.

Copyright © Great Minds PBC

2. Explain how you can use the benchmarks 0, $\frac{1}{2}$, and 1 to compare the fractions. Then write $>$, $=$, or $<$ to compare the fractions.

$$\frac{5}{8} \underline{\quad < \quad} \frac{4}{6}$$

Both fractions are 1 fractional unit more than $\frac{1}{2}$. I know that $\frac{1}{8}$ is a smaller unit than $\frac{1}{6}$, so $\frac{5}{8}$ is closer to $\frac{1}{2}$. I know $\frac{5}{8}$ is closer to $\frac{1}{2}$ than $\frac{4}{6}$, so $\frac{5}{8}$ is smaller.

I know $\frac{4}{8}$ is equivalent to $\frac{1}{2}$ and $\frac{5}{8}$ is 1 fractional unit more than $\frac{4}{8}$.

I know $\frac{3}{6}$ is equivalent to $\frac{1}{2}$ and $\frac{4}{6}$ is 1 fractional unit more than $\frac{3}{6}$.

Sixths are a larger unit than eighths, so $\frac{4}{6}$ is greater than $\frac{5}{8}$.

REMEMBER

3. Draw a quadrilateral that has 1 right angle and is not a rectangle.

Sample:

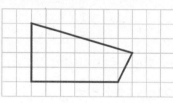

I know that a rectangle is a parallelogram that has 2 pairs of parallel sides and 4 right angles.

I see that the horizontal lines and vertical lines on the grid paper form right angles. I can trace along a horizontal line and a vertical line by using my straightedge to create a right angle.

Next, I draw 2 more sides that do not make right angles. So the shape has 4 sides, to make it a quadrilateral but not a rectangle.

         Copyright © Great Minds PBC

Draw on the place value chart to divide. Then fill in the blanks.

4. 5,264 ÷ 2

thousands	hundreds	tens	ones

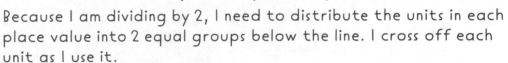

} 2,632

5,264 ÷ 2 =   2,000   +   600   +   30   +   2   =   2,632

I represent the total, 5,264, on the place value chart.

thousands	hundreds	tens	ones

Because I am dividing by 2, I need to distribute the units in each place value into 2 equal groups below the line. I cross off each unit as I use it.

I distribute the thousands. There is 1 thousand remaining, so I decompose it into 10 hundreds.

thousands	hundreds	tens	ones

Next, I distribute the hundreds, then the tens, then the ones.

Now I have 2 equal groups. I count the number of units in each place value from 1 group. I see 2 thousands 6 hundreds 3 tens 2 ones. I add the partial quotients to find the quotient.

Copyright © Great Minds PBC

Name _____ Date _____

Plot the fractions on the number line. Use >, =, or < to compare the fractions.

1. $\frac{3}{7}, \frac{4}{6}, \frac{2}{9}$

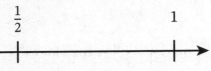

0                    $\frac{1}{2}$ 1

a. $\frac{1}{2}$ _____ $\frac{4}{6}$

b. $\frac{3}{7}$ _____ $\frac{1}{2}$

c. $\frac{4}{6}$ _____ $\frac{2}{9}$

d. $\frac{2}{9}$ _____ $\frac{3}{7}$

2. $\frac{3}{5}, \frac{2}{8}, \frac{3}{4}$

a. $\frac{3}{4}$ _____ $\frac{1}{2}$

b. $\frac{3}{5}$ _____ $\frac{3}{4}$

c. $\frac{2}{8}$ _____ $\frac{1}{2}$

d. $\frac{3}{5}$ _____ $\frac{2}{8}$

Explain how you can use the benchmarks $0, \frac{1}{2}$, and 1 to compare the fractions. Then write >, =, or < to compare the fractions.

3. $\frac{1}{7}$ _____ $\frac{1}{9}$

4. $\frac{3}{8}$ _____ $\frac{4}{6}$

REMEMBER

5. Draw a quadrilateral that has 2 right angles and is not a rectangle.

6. Draw on the place value chart to divide. Then fill in the blanks.

   4,869 ÷ 3

thousands	hundreds	tens	ones

   4,869 ÷ 3 = _____ + _____ + _____ + _____ = _____

Copyright © Great Minds PBC

Name   Date

14

1. Represent the pair of fractions with the tape diagrams. Use >, =, or < to compare the fractions.

$$\frac{2}{4} \underline{\quad < \quad} \frac{8}{12}$$

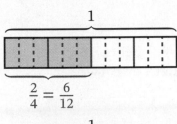

$$\frac{2}{4} = \frac{6}{12}$$

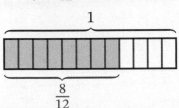

$$\frac{8}{12}$$

> I can find a **common denominator** by renaming one or both fractions so they have the same fractional unit.

I use a tape diagram to represent each fraction. I partition the first tape into fourths and shade 2 fourths. I partition the second tape into twelfths and shade 8 twelfths.

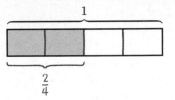

$$\frac{2}{4}$$

I can rename fourths as twelfths. In the first tape, I partition each fourth into 3 equal parts to make twelfths. There are now 6 twelfths shaded, so $\frac{2}{4} = \frac{6}{12}$.

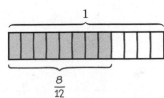

$$\frac{8}{12}$$

Now that I have like units, twelfths, it is easy to compare the fractions. More units are shaded in the second tape. I know $\frac{6}{12} < \frac{8}{12}$ so $\frac{2}{4} < \frac{8}{12}$.

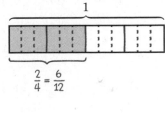

$$\frac{2}{4} = \frac{6}{12}$$

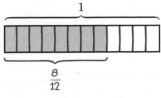

$$\frac{8}{12}$$

Copyright © Great Minds PBC

2. Represent the pair of fractions on the number line. Use >, =, or < to compare the fractions.

$\frac{4}{6}$ ___>___ $\frac{7}{12}$

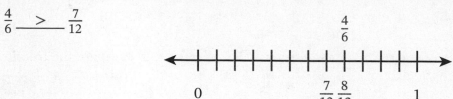

Sixths and twelfths are related. I partition the interval into sixths and plot $\frac{4}{6}$. Then I partition each sixth into two equal parts to make twelfths and plot $\frac{7}{12}$.

I can rename $\frac{4}{6}$ as twelfths.

$\frac{4}{6}$ is equivalent to $\frac{8}{12}$ because $\frac{4}{6} = \frac{2 \times 4}{2 \times 6} = \frac{8}{12}$.

I plot $\frac{8}{12}$ on the number line.

I see that $\frac{8}{12}$ is farther from 0 than $\frac{7}{12}$, so $\frac{4}{6} > \frac{7}{12}$.

3. Compare the pair of fractions by using >, =, or <. Show your thinking by using pictures, numbers, or words.

$\frac{5}{9}$ ___<___ $\frac{2}{3}$

$$\frac{2}{3} = \frac{3 \times 2}{3 \times 3} = \frac{6}{9}$$

$\frac{5}{9} < \frac{6}{9}$ so $\frac{5}{9} < \frac{2}{3}$.

I can find a common denominator for $\frac{5}{9}$ and $\frac{2}{3}$. Thirds and ninths are related units, so I rename thirds as ninths.

$\frac{2}{3}$ is equivalent to $\frac{6}{9}$ because $\frac{2}{3} = \frac{3 \times 2}{3 \times 3} = \frac{6}{9}$.

I compare $\frac{5}{9}$ and $\frac{6}{9}$. I know $\frac{5}{9}$ is less than $\frac{6}{9}$.

$\frac{2}{3}$ is equivalent to $\frac{6}{9}$ so $\frac{5}{9} < \frac{2}{3}$.

Copyright © Great Minds PBC

REMEMBER

Use the Read–Draw–Write process to solve the problem. The tape diagram has been drawn for you.

4. There are 1,950 chocolates available to put into boxes. Each box holds 9 chocolates.

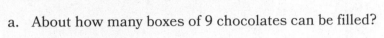
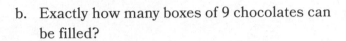

a. About how many boxes of 9 chocolates can be filled?

$1,800 \div 9 = 200$

About 200 boxes of 9 chocolates can be filled.

b. Exactly how many boxes of 9 chocolates can be filled?

$1,950 \div 9$

Quotient: 216

Remainder: 6

Exactly 216 boxes of 9 chocolates can be filled.

c. How many chocolates will not be used to fill a box?

6 chocolates will not be used to fill a box.

d. Is your answer to part (b) reasonable? Explain.

Yes, my estimate of 200 is close to 216, so my answer is reasonable.

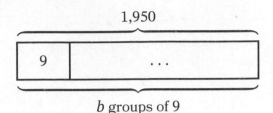

I read the problem. I read again.

I see the tape diagram and think about how it represents the problem.

I know the total is 1,950 and the size of each group is 9. The unknown is how many groups of 9 are in 1,950.

$b \times 9 = 1,950$

I estimate an answer. I know that 18 is a multiple of 9. I round 1,950 to 1,800. I can easily divide 1,800 by 9.

I divide 1,950 total chocolates into groups of 9 to find exactly how many boxes of 9 chocolates can be filled.

The quotient is 216. The remainder is 6. The quotient tells me the number of groups. The remainder tells me how many chocolates are left over and will not be put in a box.

I compare my estimate to my answer to see whether it is reasonable.

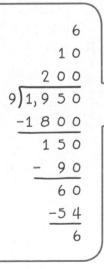

Copyright © Great Minds PBC

Name _____   Date _____

1. Represent the pair of fractions with the tape diagrams. Use >, =, or < to compare the fractions.

   $\frac{6}{10}$ _____ $\frac{2}{5}$

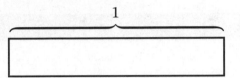

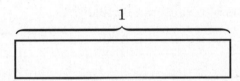

2. Represent the pair of fractions on the number line. Use >, =, or < to compare the fractions.

   $\frac{3}{4}$ _____ $\frac{10}{12}$

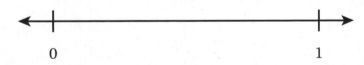

Compare each pair of fractions by using >, =, or <. Show your thinking by using pictures, numbers, or words.

3. $\frac{2}{3}$ _____ $\frac{6}{9}$ 4. $\frac{5}{12}$ _____ $\frac{1}{3}$

REMEMBER

Use the Read–Draw–Write process to solve the problem. The tape diagram has been drawn for you.

5. There are 1,778 eggs available to put into cartons. Each carton holds 6 eggs.

    a. About how many cartons of 6 eggs can be filled?

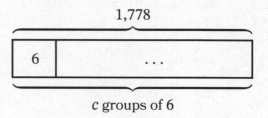

c groups of 6

    b. Exactly how many cartons of 6 eggs can be filled?

    c. How many eggs will not be used to fill a carton?

    d. Is your answer to part (b) reasonable? Explain.

Copyright © Great Minds PBC

15

Name _____ Date _____

1. Represent the pair of fractions with the tape diagrams. Use >, =, or < to compare the fractions.

$\frac{2}{3}$ __<__ $\frac{8}{10}$

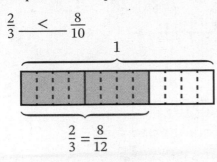

$\frac{2}{3} = \frac{8}{12}$

$\frac{8}{10}$

I can find a **common numerator** by renaming one of the fractions to create fractions with the same number of units.

I use a tape diagram to represent each fraction. I partition the first tape into thirds and shade 2 thirds. I partition the second tape into tenths and shade 8 tenths.

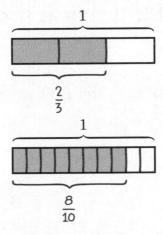

$\frac{2}{3}$

$\frac{8}{10}$

I can rename $\frac{2}{3}$ as $\frac{8}{12}$ because $\frac{2}{3} = \frac{4 \times 2}{4 \times 3} = \frac{8}{12}$. Now both fractions have the same numerator. In the first tape, I partition each third into 4 parts to make twelfths. There are now 8 twelfths shaded.

Tenths are larger fractional units than twelfths, so $\frac{8}{12} < \frac{8}{10}$.

That means that $\frac{2}{3} < \frac{8}{10}$.

Copyright © Great Minds PBC

2. Represent the pair of fractions on the number line. Use >, =, or < to compare the fractions.

$\frac{3}{4}$ __=__ $\frac{9}{12}$

I partition the interval into fourths and plot $\frac{3}{4}$.

Then I partition each fourth into three equal parts to make twelfths and plot $\frac{9}{12}$.

Both fractions name the same point on the number line. So $\frac{3}{4} = \frac{9}{12}$.

Copyright © Great Minds PBC

3. Compare the pair of fractions by using >, =, or <. Show your thinking by using pictures, numbers, or words.

$\frac{2}{5}$ __<__ $\frac{4}{6}$

$$\frac{2}{5} = \frac{2 \times 2}{2 \times 5} = \frac{4}{10}$$

$$\frac{4}{10} < \frac{4}{6}$$

$$\frac{2}{5} < \frac{4}{6}$$

I can find a common numerator for $\frac{2}{5}$ and $\frac{4}{6}$. I see that the numerators are related. 4 is 2 times as much as 2. I can multiply the numerator and the denominator of $\frac{2}{5}$ by 2.

$$\frac{2}{5} = \frac{2 \times 2}{2 \times 5} = \frac{4}{4}$$

Now I can compare $\frac{4}{10}$ and $\frac{4}{6}$.

The fractions have the same number of units in the numerator.

I can compare the sizes of the fractional units in the denominators. I know that tenths are smaller than sixths, so $\frac{4}{10} < \frac{4}{6}$.

Because $\frac{2}{5}$ and $\frac{4}{10}$ are equivalent, I know that $\frac{2}{5} < \frac{4}{6}$.

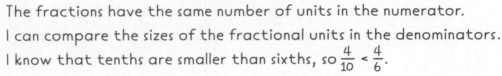

REMEMBER

4. Write two different equations to show $\frac{6}{5}$ decomposed as a sum of fractions. Draw a model to justify one of your answers.

Sample:

$\frac{6}{5} = \underline{\frac{4}{5}} + \underline{\frac{2}{5}}$

$\frac{6}{5} = \underline{\frac{5}{5}} + \underline{\frac{1}{5}}$

I can draw a tape diagram with 6 total units.

Each unit is $\frac{1}{5}$.

$\frac{6}{5}$ is the total, and $\frac{5}{5}$ and $\frac{1}{5}$ are the parts.

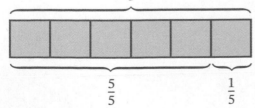

$\frac{6}{5}$

$\frac{5}{5}$ $\frac{1}{5}$

Copyright © Great Minds PBC

Name _____   Date _____

1. Represent the pair of fractions with the tape diagrams. Use >, =, or < to compare the fractions.

$\dfrac{4}{8}$ _____ $\dfrac{2}{3}$

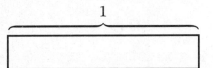

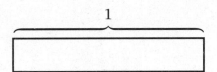

2. Represent the fractions on the number line. Use >, =, or < to compare the fractions.

$\dfrac{4}{5}$ _____ $\dfrac{8}{10}$

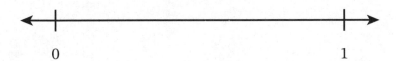

Compare each pair of fractions by using >, =, or <. Show your thinking by using pictures, numbers, or words.

3. $\dfrac{2}{5}$ _____ $\dfrac{4}{8}$

4. $\dfrac{3}{4}$ _____ $\dfrac{6}{10}$

REMEMBER

5. Write two different equations to show $\frac{8}{6}$ decomposed as a sum of fractions. Draw a model to justify one of your answers.

$\frac{8}{6} = $ _____ $+$ _____

$\frac{8}{6} = $ _____ $+$ _____

Copyright © Great Minds PBC

16

Name _____ Date _____

1. Rename the pair of fractions by using common denominators. Use >, =, or < to compare the fractions.

$$\frac{4}{5} \underline{\quad > \quad} \frac{2}{3}$$

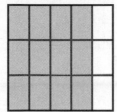 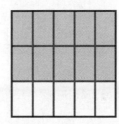

$$\frac{4}{5} = \frac{4 \times 3}{5 \times 3} = \frac{12}{15} \qquad \frac{2}{3} = \frac{2 \times 5}{3 \times 5} = \frac{10}{15}$$

I use multiplication to find a common denominator, fifteenths.

I can represent both fractions by using area models. I shade $\frac{4}{5}$ vertically in one area model. I shade $\frac{2}{3}$ horizontally in another area model.

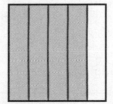 

The first area model has 5 vertical sections. The second area model has 3 horizontal sections. I partition the fifths in the first area model into 3 equal parts to make fifteenths. I partition the thirds in the second area model into 5 equal parts to make fifteenths.

The area models represent $\frac{12}{15}$ and $\frac{10}{15}$.
$\frac{12}{15} > \frac{10}{15}$, so $\frac{4}{5} > \frac{2}{3}$.

2. Rename the pair of fractions by using common numerators. Use >, =, or < to compare the fractions.

$$\frac{5}{12} \underline{\quad < \quad} \frac{3}{7}$$

$$\frac{5}{12} = \frac{3 \times 5}{3 \times 12} = \frac{15}{36}$$

$$\frac{3}{7} = \frac{5 \times 3}{5 \times 7} = \frac{15}{35}$$

$$\frac{15}{36} < \frac{15}{35}$$

$$\frac{5}{12} < \frac{3}{7}$$

The numerators 5 and 3 are not related. I can use the numerators as factors to find a common numerator, 5 × 3 = 15.

I multiply the numerator and the denominator of $\frac{5}{12}$ by 3 to get a numerator of 15.

I multiply the numerator and the denominator of $\frac{3}{7}$ by 5 to get a numerator of 15.

$\frac{15}{36} < \frac{15}{35}$ because thirty-sixths are smaller fractional units than thirty-fifths. So $\frac{5}{12} < \frac{3}{7}$.

Copyright © Great Minds PBC

REMEMBER

3. The crew on a ferry keeps track of how many passengers the ferry has each day.

    The ferry has 10,164 passengers on Monday and 6,943 passengers on Tuesday.

    On Wednesday, the ferry has 2,304 more passengers than on Monday.

    On Thursday, the ferry has 1,320 fewer passengers than on Tuesday.

    a. How many passengers does the ferry have in all? Use equations to show how you found your answer.

    $$10,164 + 2,304 = 12,468$$

    $$6,943 - 1,320 = 5,623$$

    $$10,164 + 6,943 + 12,468 + 5,623 = 35,198$$

    The ferry has 35,198 passengers in all.

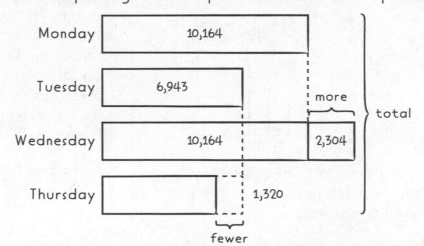

> I draw a tape diagram to help me make sense of the problem.
>
> | Monday | 10,164 |
> | Tuesday | 6,943 |
> | Wednesday | 10,164 | 2,304 |
> | Thursday | | 1,320 |
>
> more
> fewer
> total

To find the number of passengers on Wednesday I add 10,164 and 2,304.

To find the number of passengers on Thursday I subtract 1,320 from 6,943.

I add each day's total number of passengers to find how many passengers there are in all.

Copyright © Great Minds PBC

b. Is your answer reasonable? Explain.

Yes. I estimated to see if my answer is reasonable by rounding each number to the nearest thousand.

$$10{,}000 + 2{,}000 = 12{,}000$$

$$7{,}000 - 1{,}000 = 6{,}000$$

$$10{,}000 + 7{,}000 + 12{,}000 + 6{,}000 = 35{,}000$$

My estimate, 35,000, is close to the actual answer, 35,198, so my answer is reasonable.

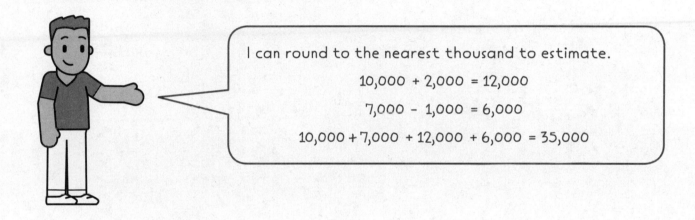

I can round to the nearest thousand to estimate.
10,000 + 2,000 = 12,000
7,000 − 1,000 = 6,000
10,000 + 7,000 + 12,000 + 6,000 = 35,000

16

Name _____ Date _____

Rename each pair of fractions by using common denominators. Use >, =, or < to compare the fractions.

1. $\frac{4}{6}$ _____ $\frac{3}{4}$

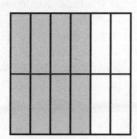 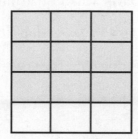

2. $\frac{1}{2}$ _____ $\frac{4}{5}$

Rename each pair of fractions by using common numerators. Use >, =, or < to compare the fractions.

3. $\frac{4}{7}$ _____ $\frac{5}{9}$

4. $\frac{2}{3}$ _____ $\frac{8}{10}$

Copyright © Great Minds PBC

REMEMBER

5. Mr. Davis keeps track of how many steps he takes each day.

   - He walks 11,234 steps on Monday and 7,586 steps on Tuesday.

   - On Wednesday, Mr. Davis walks 1,620 more steps than on Monday.

   - On Thursday, he walks 430 fewer steps than on Tuesday.

   a. How many total steps does Mr. Davis walk? Use equations to show how you found your answer.

   b. Is your answer reasonable? Explain.

Copyright © Great Minds PBC

_____   _____
Name   Date

Compare the fractions by using >, =, or <. Explain your strategy by using pictures, numbers, or words.

1. $\dfrac{17}{5}$ __>__ $\dfrac{15}{6}$

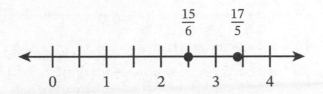

I can plot the fractions on a number line to compare them to benchmark numbers.

I know $\dfrac{17}{5}$ is between 3 and 4 because $\dfrac{15}{5}$ = 3 and $\dfrac{20}{5}$ = 4. I plot $\dfrac{17}{5}$ as a little less than $3\dfrac{1}{2}$.

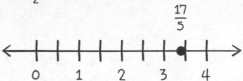

I know $\dfrac{15}{6}$ is between 2 and 3 because $\dfrac{12}{6}$ = 2 and $\dfrac{18}{6}$ = 3. I plot $\dfrac{15}{6}$ as halfway between 2 and 3.

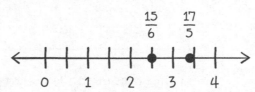

I see from the number line that $\dfrac{17}{5} > \dfrac{15}{6}$.

Copyright © Great Minds PBC

2. $4\frac{7}{12}$ ___<___ $4\frac{3}{4}$

The whole number parts of both numbers are equal, so I can compare the fractional parts.

I know that $\frac{3}{4} = \frac{3 \times 3}{3 \times 4} = \frac{9}{12}$ and $\frac{7}{12} < \frac{9}{12}$. Because $\frac{7}{12} < \frac{9}{12}$, that means $4\frac{7}{12} < 4\frac{3}{4}$.

I find a common denominator of $\frac{7}{12}$ and $\frac{3}{4}$. Because 4 is a factor of 12, a common denominator of the two fractions is 12.

$$\frac{3}{4} = \frac{3 \times 3}{3 \times 4} = \frac{9}{12}$$

I compare the units. I see that $\frac{7}{12} < \frac{9}{12}$, so $4\frac{7}{12} < 4\frac{3}{4}$.

3. $\frac{5}{2}$ ___<___ $\frac{10}{3}$

$$\frac{5}{2} = \frac{2 \times 5}{2 \times 2} = \frac{10}{4}$$
$$\frac{10}{4} < \frac{10}{3}$$
$$\frac{5}{2} < \frac{10}{3}$$

I find a common numerator for $\frac{5}{2}$ and $\frac{10}{3}$. Because 5 is a factor of 10, a common numerator for the two fractions is 10.

$$\frac{5}{2} = \frac{2 \times 5}{2 \times 2} = \frac{10}{4}$$

I compare the fractional units. Fourths is a smaller unit than thirds, so $\frac{10}{4} < \frac{10}{3}$.

So $\frac{5}{2} < \frac{10}{3}$.

                                    Copyright © Great Minds PBC

REMEMBER

Add by using the standard algorithm.

4.

	2,	3	2	9
+	6,	0	7₁	6₁
	8,	4	0	5

First, I add the ones column,
9 ones + 6 ones = 15 ones. I rename 15 ones as
1 ten 5 ones.

	2,	3	2	⑨
+	6,	0	7	⑥
			1	⑤

I add the tens column,
2 tens + 7 tens + 1 ten = 10 tens. I rename 10 tens as
1 hundred 0 tens.

	2,	3	②	9
+	6,	0	⑦	6
			1 ⓪	5

I add the hundreds column,
3 hundreds + 0 hundreds + 1 hundred = 4 hundreds.

	2,	③	2	9
+	6,	⓪	7	6
		④	0	5

I add the thousands column,
2 thousands + 6 thousands = 8 thousands.

	②,	3	2	9
+	⑥,	0	7	6
	⑧,	4	0	5

2,329 + 6,076 = 8,405

My answer is reasonable because 8,405 is close to my
estimate of 8,000.

I round each number to
its largest place value to
estimate the sum.

$2,329 \approx 2,000$

$6,076 \approx 6,000$

2 thousands + 6 thousands
= 8 thousands

So my estimate is
8 thousands, or 8,000.

Name _____   Date _____

Compare the fractions by using >, =, or <. Explain your strategy by using pictures, numbers, or words.

1. $\dfrac{14}{4}$ _____ $\dfrac{12}{5}$

2. $3\dfrac{2}{3}$ _____ $3\dfrac{3}{6}$

3. $\dfrac{11}{3}$ _____ $\dfrac{7}{2}$

4. $2\dfrac{2}{4}$ _____ $2\dfrac{6}{10}$

Copyright © Great Minds PBC

REMEMBER

5. Add by using the standard algorithm.

	4	8	2,	3	0	5
+		5	3,	7	4	6

Copyright © Great Minds PBC

FAMILY MATH
Adding and Subtracting Fractions

Dear Family,

Your student is learning to add and subtract fractions with like units. They use unit form, fraction form, and number lines to support their thinking. They learn to break apart a whole number so they can easily subtract. Your student also uses drawings to help them determine whether they need to add or subtract when solving word problems. They estimate to decide whether their answer is reasonable. The skills your student is learning now will support them later when they add and subtract mixed numbers.

Adding and Subtracting Fractions with Like Units

- Unit form

  4 eighths + 3 eighths = 7 eighths

  8 tenths − 6 tenths = 2 tenths

- Fraction form

  $$\frac{5}{10} + \frac{2}{10} = \frac{7}{10}$$

  $$\frac{6}{8} - \frac{4}{8} = \frac{2}{8}$$

$$2 - \frac{7}{10} = 1\frac{3}{10}$$

$1 \qquad \frac{10}{10}$

$$\frac{10}{10} - \frac{7}{10} = \frac{3}{10}$$

$$1 + \frac{3}{10} = 1\frac{3}{10}$$

Students use unit form and fraction form to write equations. Unit form helps students see that adding and subtracting fractions is similar to adding and subtracting whole numbers. For example, 4 ones + 3 ones = 7 ones so 4 eighths + 3 eighths = 7 eighths.

Students learn to break apart a whole number by using fractions. 2 is broken into 1 and $\frac{10}{10}$ because tenths is the unit that is subtracted from 2.

$\frac{11}{12}$

$a \qquad \frac{7}{12}$

$$\frac{11}{12} - \frac{7}{12} = a$$

$$\frac{4}{12} = a$$

There is $\frac{4}{12}$ of a pan of brownies left.

Drawing a model, such as a tape diagram, helps students decide whether to use addition or subtraction to solve a word problem.

Copyright © Great Minds PBC

At-Home Activity

Pizza Fractions

Practice adding and subtracting fractions with your student when eating pizza or any other food cut into equal pieces. You may also draw a picture to represent the pizza. Ask them questions such as the following to guide their thinking.

- "How many slices of pizza are in the whole? What fraction can we use to describe 1 slice?"

- "What fraction can we use to describe the whole pizza?"

- "What addition problem represents the fraction of pizza eaten if you eat 2 slices?"

- "What subtraction problem represents the fraction of pizza left over after we eat 3 slices of the whole pizza? What if we eat 5 slices?"

Copyright © Great Minds PBC

_____   _____
Name   Date

Estimate the sum or difference.

1. $\frac{6}{10} + \frac{14}{10} \approx 2$

I think about equivalent fractions and the benchmarks $\frac{1}{2}$ and 1 to help me estimate.

I know $\frac{5}{10} = \frac{1}{2}$ so $\frac{6}{10}$ is close to $\frac{5}{10}$.

$$\frac{6}{10} \approx \frac{1}{2}$$

I use a number bond to decompose $\frac{14}{10}$.

$\frac{10}{10} = 1$ and $\frac{4}{10}$ is about $\frac{1}{2}$.

$$\frac{14}{10} \approx 1\frac{1}{2}$$

```
      10        4
      10       10
```

I find the sum to make my estimate.

$$\frac{1}{2} + 1\frac{1}{2} = 2$$

2. $\frac{5}{6} - \frac{2}{6} \approx \frac{1}{2}$

I can draw a number line to estimate.

I label the benchmarks $0, \frac{1}{2},$ and 1.

Then I label the equivalent fractions in sixths.

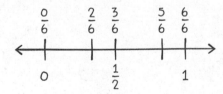

$\frac{5}{6}$ is close to 1 and $\frac{2}{6}$ is close to $\frac{1}{2}$.

I subtract to make my estimate.

$$1 - \frac{1}{2} = \frac{1}{2}$$

Use the Read–Draw–Write process to solve the problem.

3. A nail is $\frac{9}{8}$ inches long. A screw is $\frac{5}{8}$ inches long. About how much longer is the nail than the screw?

$\frac{9}{8} - \frac{5}{8} \approx \frac{1}{2}$

The nail is about $\frac{1}{2}$ inch longer than the screw.

I read the problem. I read again.

As I reread, I think about what I can draw.

I draw a tape diagram. I draw one tape to show the length of the nail. I draw another tape to show the length of the screw.

The difference between the 2 lengths is unknown.

I use benchmarks to estimate.

$\frac{9}{8}$ is close to $\frac{8}{8}$, so $\frac{9}{8} \approx 1$.

$\frac{5}{8}$ is close to $\frac{4}{8}$, so $\frac{5}{8} \approx \frac{1}{2}$.

I subtract to make my estimate.

$1 - \frac{1}{2} = \frac{1}{2}$

Nail $\boxed{\frac{9}{8}}$

Screw $\boxed{\frac{5}{8}}$

Copyright © Great Minds PBC

REMEMBER

4. Show that the fractions are equivalent by drawing an area model. Then express the equivalence by using multiplication.

$\frac{2}{5}$ and $\frac{4}{10}$

$$\frac{2}{5} = \frac{2 \times 2}{2 \times 5} = \frac{4}{10}$$

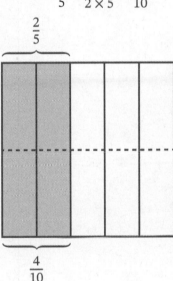

For $\frac{2}{5}$, I draw an area model partitioned into fifths and shade 2 equal parts.

To show $\frac{4}{10}$, I can draw a horizontal line across the middle of the area model.

Now the area model is partitioned into 10 equal parts with 4 parts shaded.

I can see that $\frac{2}{5}$ is the same as $\frac{4}{10}$.

The area model shows each fifth partitioned into 2 equal parts to make tenths. Tenths have twice as many units as fifths.

I multiply the numerator and the denominator by 2 to show 2 times as many shaded parts and 2 times as many total parts.

Copyright © Great Minds PBC

Name _____   Date _____

Estimate the sum or difference.

1. $\frac{4}{10} + \frac{9}{10} \approx$

2. $\frac{3}{5} + \frac{11}{5} + \frac{6}{5} \approx$

3. $\frac{11}{12} - \frac{7}{12} \approx$

4. $\frac{13}{6} - \frac{4}{6} \approx$

Use the Read–Draw–Write process to solve each problem.

5. On Saturday, it rained $\frac{19}{10}$ inches. The record for the most rain on that date was $\frac{11}{10}$ inches. About how many more inches than the record did it rain on Saturday?

6. Liz is mixing bird seed for her bird feeder. She uses $\frac{7}{8}$ cups of sunflower seeds, $\frac{3}{8}$ cups of corn, and $\frac{9}{8}$ cups of peanuts. About how many cups does Liz mix altogether?

Copyright © Great Minds PBC

REMEMBER

7. Show that the fractions are equivalent by drawing an area model. Then express the equivalence by using multiplication.

$\frac{1}{5}$ and $\frac{2}{10}$

Copyright © Great Minds PBC

19

Name _____ Date _____

1. Add. Write the sum in unit form.

7 twelfths + 3 twelfths = __10 twelfths__

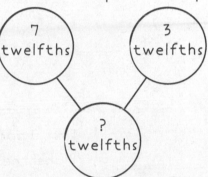

I can use a number bond to represent the problem.

7 twelfths

3 twelfths

? twelfths

I know that 7 twelfths + 3 twelfths is equal to some twelfths.

I know 7 + 3 = 10 so 7 twelfths + 3 twelfths = 10 twelfths.

2. Subtract. Write the difference in fraction form.

$\frac{7}{12} - \frac{2}{12} =$ __$\frac{5}{12}$__

The fractions have like units of twelfths.

I know 7 − 2 = 5 so
7 twelfths − 2 twelfths = 5 twelfths.

3. Subtract. Write the difference in fraction form. Use the number line to represent the subtraction.

$$\frac{8}{10} - \frac{7}{10} = \underline{\frac{1}{10}}$$

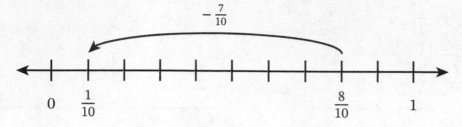

First, I partition the interval into tenths. I label the total, which is $\frac{8}{10}$.

Now I show the subtraction of $\frac{7}{10}$ from the total with an arrow. The arrow shows the difference on the number line.

Copyright © Great Minds PBC

REMEMBER

4. Complete the equation to show an equivalent fraction. You may draw an area model to help you.

$$\frac{4}{12} = \frac{1}{3}$$

I draw an area model that shows $\frac{4}{12}$.

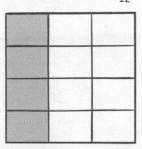

I know that the numerator of the equivalent fraction is 1. That means the equivalent fraction has 1 unit.

I circle to compose 4 units in the area model.

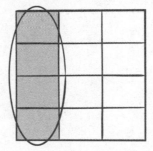

I can represent the composition with division.

$$4 \div 4 = 1$$

I divide the numerator and the denominator of $\frac{4}{12}$ by 4 to create an equivalent number of thirds.

$$\frac{4}{12} = \frac{4 \div 4}{12 \div 4} = \frac{1}{3}$$

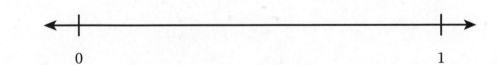

19

Name _____ Date _____

Add or subtract. Write the sum or difference in unit form.

1. 4 tenths + 4 tenths = _____

2. 6 thirds − 5 thirds = _____

Add or subtract. Write the sum or difference in fraction form.

3. $\frac{5}{8} + \frac{6}{8} =$ _____

4. $\frac{4}{4} - \frac{3}{4} =$ _____

Add or subtract. Write the sum or difference in fraction form. Use the number line to represent the addition or subtraction.

5. $\frac{2}{6} + \frac{3}{6} =$ _____

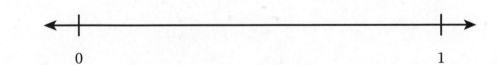

0   1

6. $\frac{6}{8} - \frac{4}{8} =$ _____

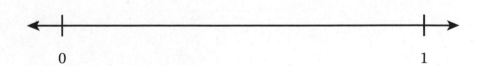

0   1

Copyright © Great Minds PBC

REMEMBER

7. Complete the equation to show an equivalent fraction. You may draw an area model to help you.

$$\frac{8}{10} = \frac{4}{\boxed{}}$$

Copyright © Great Minds PBC

20

Name _____   Date _____

1. Subtract. Use a number line to represent the subtraction.

$$1 - \frac{7}{12} = \underline{\frac{5}{12}}$$

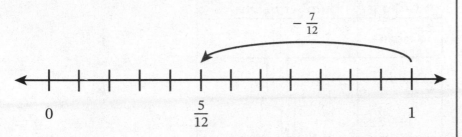

$$-\frac{7}{12}$$

0          $\frac{5}{12}$ 1

> I need to subtract $\frac{7}{12}$ from 1. So I partition the number line into twelfths.
>
> Then I show the subtraction of $\frac{7}{12}$ from 1.

Rename or decompose the total to subtract.

2. $1 - \frac{2}{8} = \underline{\frac{6}{8}}$

$$\frac{8}{8} - \frac{2}{8} = \frac{6}{8}$$

> I can rename 1 as $\frac{8}{8}$. Then I subtract the part, $\frac{2}{8}$, from $\frac{8}{8}$.

> I can decompose 3 into 2 and 1. Then I rename 1 as $\frac{10}{10}$.
>
> 3
> ╱ ╲
> 2 $1 = \frac{10}{10}$
>
> Next I can subtract $\frac{8}{10}$ from $\frac{10}{10}$ to get $\frac{2}{10}$. Then I add 2 to $\frac{2}{10}$.
>
> $2 + \frac{2}{10} = 2\frac{2}{10}$

3. $3 - \frac{8}{10} = \underline{2\frac{2}{10}}$

$$\begin{array}{c} \\ \diagup \diagdown \\ 2 \quad \frac{10}{10} \end{array}$$

Copyright © Great Minds PBC

REMEMBER

4. Liz collects leaves in her backyard. The tally chart shows the number of leaves she collects in each color. Use the data in the chart to draw a scaled picture graph.

Liz's Leaf Collection	
Color	**Number of Leaves**
Green	||
Yellow	||||
Red	‖‖|
Brown	||
Orange	‖‖ |||

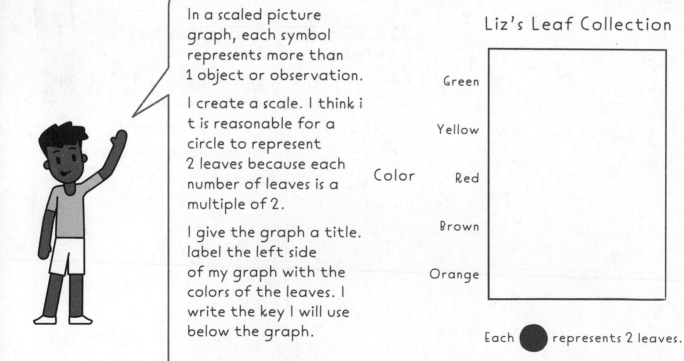

In a scaled picture graph, each symbol represents more than 1 object or observation.

I create a scale. I think it is reasonable for a circle to represent 2 leaves because each number of leaves is a multiple of 2.

I give the graph a title. label the left side of my graph with the colors of the leaves. I write the key I will use below the graph.

Liz's Leaf Collection

Color

Green

Yellow

Red

Brown

Orange

Each ● represents 2 leaves.

Copyright © Great Minds PBC

Liz's Leaf Collection

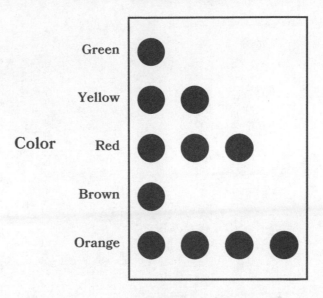

Color

Each ● represents 2 leaves.

Now I can draw circles to represent my data.

I draw the following circles.

- 1 circle to represent 2 green leaves.

- 2 circles to represent 4 yellow leaves because $2 + 2 = 4$.

- 3 circles to represent 6 red leaves because $2 + 2 + 2 = 6$.

- 1 circle to represent 2 brown leaves.

- 4 circles to represent 8 orange leaves because $2 + 2 + 2 + 2 = 8$.

Name _____   Date _____

Subtract. Use a number line to represent the subtraction.

1. $1 - \frac{2}{8} = $ _____

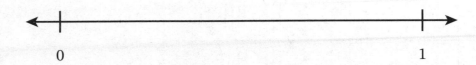

2. $2 - \frac{5}{6} = $ _____

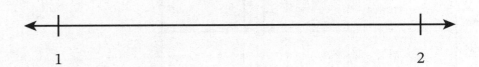

Rename or decompose the total to subtract.

3. $1 - \frac{8}{12} = $ _____

4. $2 - \frac{4}{6} = $ _____

Copyright © Great Minds PBC

REMEMBER

5. Miss Wong surveys her students about their favorite fruits. The tally chart shows the results of the survey. Use the data in the chart to draw a scaled picture graph.

Students' Favorite Fruits	
Fruit	**Number of Students**
Peach	卌 |||
Pear	卌	
Apple	||
Watermelon	||||
Banana	||||

Copyright © Great Minds PBC

Name _____   Date _____

Use the Read–Draw–Write process to solve the problem.

1. Deepa jogs $\frac{4}{10}$ miles on Monday, $\frac{6}{10}$ miles on Tuesday, and $\frac{9}{10}$ miles on Thursday.

   a. How many miles does Deepa jog altogether?

      $$\frac{4}{10} + \frac{6}{10} + \frac{9}{10} = \frac{19}{10} = 1\frac{9}{10}$$

   b. Is your answer from part (a) reasonable? Explain.

      My answer is reasonable because $\frac{4}{10}$ is close to $\frac{1}{2}$, $\frac{6}{10}$ is close to $\frac{1}{2}$, and $\frac{9}{10}$ is close to 1.

      So $\frac{1}{2} + \frac{1}{2} + 1 = 2$, which is close to my answer of $1\frac{9}{10}$.

I read the problem. I read again.

As I reread, I think about what I can draw.

I draw a tape diagram to show that Deepa jogs $\frac{4}{10}$ miles on Monday, $\frac{6}{10}$ miles on Tuesday, and $\frac{9}{10}$ miles on Thursday.

The unknown is how many miles she jogs altogether. I can add the fractions to find the total.

The sum is a fraction greater than 1, $\frac{19}{10}$. I can rename it as a mixed number, $1\frac{9}{10}$. Deepa jogs $1\frac{9}{10}$ miles altogether.

I can check to see whether my answer is reasonable. $\frac{4}{10}$ is close to $\frac{1}{2}$, $\frac{6}{10}$ is close to $\frac{1}{2}$, and $\frac{9}{10}$ is close to 1.

$$\frac{1}{2} + \frac{1}{2} + 1 = 2$$

Deepa jogs about 2 miles.

$1\frac{9}{10}$ is close to 2, so my answer is reasonable.

REMEMBER

2. Measure the lengths of the fish to the nearest half inch. Create a line plot to represent your data.

Copyright © Great Minds PBC

Lengths of Fish

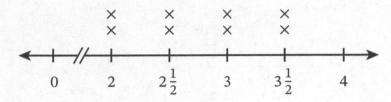

Length (inches)

I measure each fish. I line up the tick mark for zero on my ruler with the end of the fish. I measure to the nearest half inch.

The shortest measurement is 2 inches, and the longest measurement is $3\frac{1}{2}$ inches. Because I measure to the nearest half inch, I make intervals of $\frac{1}{2}$ on the line plot.

First, I draw tick marks for 2, 3, and 4. Then I draw tick marks for $2\frac{1}{2}$ and $3\frac{1}{2}$.

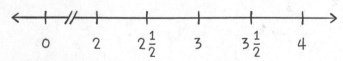

I make an X above the length on the line plot to represent each fish I measured.

_____   _____
Name   Date

Use the Read–Draw–Write process to solve each problem.

1. Mia rides her bike $\frac{4}{10}$ miles to school and $\frac{3}{10}$ miles to her friend's house. How many total miles does she ride her bike?

2. A bag is $\frac{7}{8}$ full of chicken feed. Oka pours $\frac{4}{8}$ of the bag into the chicken feeder. How full is the bag now?

3. Ivan buys $\frac{1}{4}$ pound of apples, $\frac{3}{4}$ pounds of oranges, and $\frac{3}{4}$ pounds of bananas.

    a. How many pounds of fruit does Ivan buy?

    b. Is your answer from part (a) reasonable? Explain.

Copyright © Great Minds PBC

REMEMBER

5. Measure the lengths of the keys to the nearest half inch. Create a line plot to represent your data.

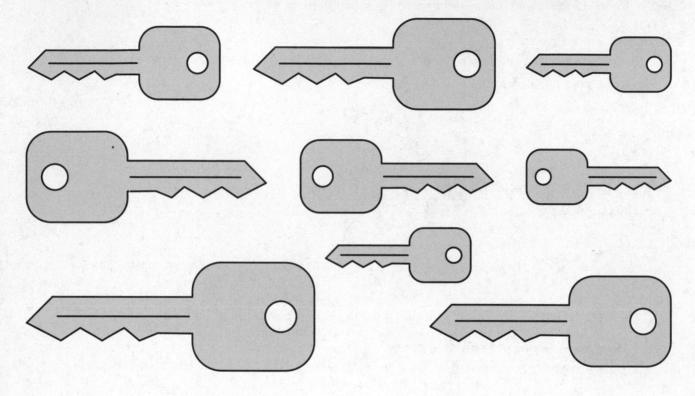

Copyright © Great Minds PBC

Name _____   Date _____

1. Represent the addition by using a number line. Then complete the statement.

$$\frac{2}{3} + \frac{2}{6} = \boxed{\frac{4}{6}} + \frac{2}{6} = \boxed{\frac{6}{6}}$$

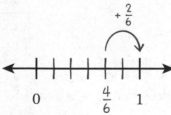

I can use the number line to rename thirds as sixths.

First, I partition the number line into thirds.

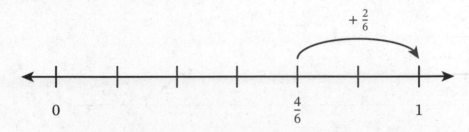

Then I decompose each third into 2 equal parts to make sixths. I see that $\frac{2}{3}$ is equivalent to $\frac{4}{6}$.

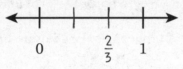

Now I show the addition on the number line.

$$\frac{2}{3} + \frac{2}{6} = \frac{4}{6} + \frac{2}{6} = \frac{6}{6}$$

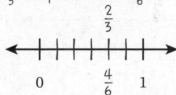

So the sum is 1, or $\frac{6}{6}$.

2. Rename to find the sum. Show your work.

$\frac{1}{3} + \frac{5}{12} = \frac{4}{12} + \frac{5}{12} = \frac{9}{12}$

> I need like units to add, so I rename $\frac{1}{3}$ as twelfths. I can use multiplication to make an equivalent fraction. I see there are 4 twelfths in $\frac{1}{3}$.
>
> $$\frac{1}{3} = \frac{4 \times 1}{4 \times 3} = \frac{4}{12}$$
>
> Now I add.
>
> $$\frac{4}{12} + \frac{5}{12} = \frac{9}{12}$$

REMEMBER

Use the Read–Draw–Write process to solve the problem.

3. Mr. Davis buys 3 crates of tangerines. Each crate has 36 tangerines.

   How many tangerines does he buy?

   Mr. Davis buys 108 tangerines.

> I read the problem. I read again.
>
> As I reread, I think about what I can draw.
>
> I draw a tape diagram with 3 parts to represent the crates of tangerines.
>
> ?
> | 36 | 36 | 36 |
>
> I see 3 groups of 36. The unknown is the total. I can decompose 36 into 3 tens 6 ones to help me multiply.
>
> $$3 \times 36 = (3 \times 30) + (3 \times 6)$$
> $$= 90 + 18$$
> $$= 108$$

Copyright © Great Minds PBC

Name

Date

Represent the addition by using a number line. Then complete the statements.

1. $\frac{2}{12} + \frac{2}{6} = \frac{2}{12} + \frac{\boxed{}}{\boxed{}} = \frac{\boxed{}}{\boxed{}}$

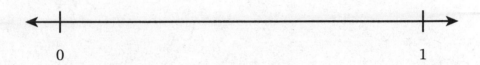

```
<---+-------------------+--->
    0                   1
```

2. $\frac{1}{10} + \frac{1}{2} = \frac{\boxed{}}{\boxed{}} + \frac{\boxed{}}{\boxed{}} = \frac{\boxed{}}{\boxed{}}$

```
<---+-------------------+--->
    0                   1
```

Rename to find the sum. Show your work.

3. $\frac{2}{3} + \frac{1}{6} = \frac{\boxed{}}{\boxed{}} + \frac{\boxed{}}{\boxed{}} = \frac{\boxed{}}{\boxed{}}$

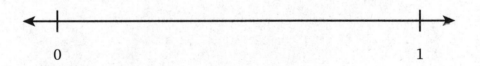

4. $\frac{3}{4} + \frac{1}{12} =$

5. $\frac{3}{8} + \frac{1}{2} =$

Copyright © Great Minds PBC

REMEMBER

Use the Read–Draw–Write process to solve the problem.

6. Miss Diaz buys 4 bags of lemons. Each bag has 46 lemons. How many lemons does Miss Diaz buy?

Copyright © Great Minds PBC

FAMILY MATH
Adding and Subtracting Mixed Numbers

Dear Family,

Your student is using a variety of familiar addition and subtraction strategies to add and subtract mixed numbers. They draw and use models and write equations to solve word problems. They apply their understanding of adding and subtracting mixed numbers to answer questions about the data in a line plot. Then they use given data to make a line plot and write their own questions.

$$5\frac{2}{4} + \frac{3}{4} = \underline{\quad 6\frac{1}{4} \quad}$$

$$5\frac{2}{4} \xrightarrow{+\frac{2}{4}} 6 \xrightarrow{+\frac{1}{4}} 6\frac{1}{4}$$

$$5\frac{2}{4} + \frac{3}{4} = \underline{5\frac{1}{4} + 1} = 6\frac{1}{4}$$

$$\overset{\displaystyle 5\frac{1}{4} \quad \frac{1}{4}}{\wedge}$$

Students can use different break apart strategies to add a fraction to a mixed number.

$$1\frac{1}{5} - \frac{4}{5} = \underline{\quad \frac{2}{5} \quad}$$

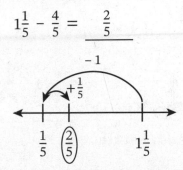

Students can subtract with mixed numbers by using a number line to help them think about the numbers.

Copyright © Great Minds PBC

Weights of Bags of Apples Sold on Monday

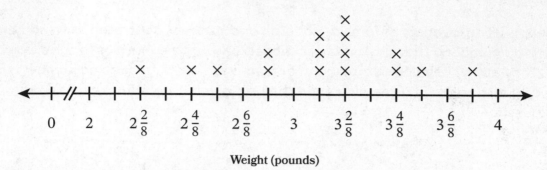

Weight (pounds)

Students analyze a line plot. They answer questions related to the line plot which also includes deciding whether a claim about the data is true or false.

At-Home Activity

Measure, Add, Subtract

Ask your student to measure and record the lengths of two household objects, such as crayons, books, or shoelaces. Have them measure the items to the nearest eighth of an inch. Have them add the lengths together to determine the total length of both objects. Then have them subtract to find how much shorter one object is than the other. Encourage your student to explain their thinking to you.

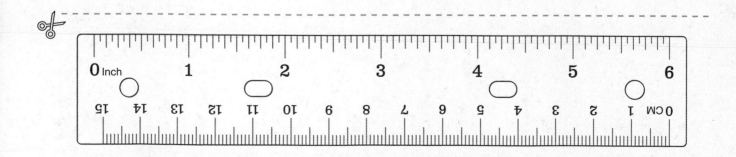

                    Copyright © Great Minds PBC

23

Name _____ Date _____

Add. Show your thinking.

1. $5\frac{4}{9} + \frac{8}{9} =$ _____ $6\frac{3}{9}$

$$5\frac{4}{9} + \frac{8}{9} = 5\frac{3}{9} + 1 = 6\frac{3}{9}$$

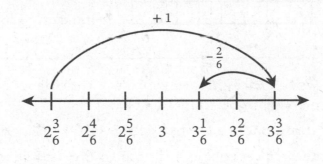

$5\frac{3}{9}$ $\frac{1}{9}$

I can draw a number bond to help me solve the problem.

I can decompose $5\frac{4}{9}$ into $5\frac{3}{9}$ and $\frac{1}{9}$ because I know $\frac{1}{9}$ and $\frac{8}{9}$ make 1.

I add $5\frac{3}{9}$ and 1 to get $6\frac{3}{9}$.

2. $\frac{4}{6} + 2\frac{3}{6} =$ _____ $3\frac{1}{6}$

$+1$

$-\frac{2}{6}$

$2\frac{3}{6}$ $2\frac{4}{6}$ $2\frac{5}{6}$ 3 $3\frac{1}{6}$ $3\frac{2}{6}$ $3\frac{3}{6}$

I can use compensation and a number line to help me solve the problem.

I start at $2\frac{3}{6}$ and add 1 because it is a benchmark number close to $\frac{4}{6}$.

I added $\frac{2}{6}$ more than I needed, so I subtract $\frac{2}{6}$ from $3\frac{3}{6}$.

So $\frac{4}{6} + 2\frac{3}{6} = 3\frac{1}{6}$.

Copyright © Great Minds PBC

REMEMBER

3. Draw and shade 2 different rectangles that each have an area of 36 square units.
    Show how to find the area and perimeter of each rectangle. Sample:

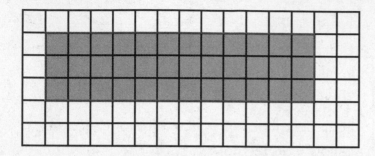

Equation to find area: ___3 × 12 = 36___

Area: ___36___ square units

Equation to find perimeter:
___(2 × 3) + (2 × 12) = 6 + 24 = 30___

Perimeter: ___30___ units

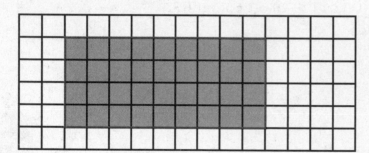

Equation to find area: ___4 × 9 = 36___

Area: ___36___ square units

Equation to find perimeter:
___(2 × 4) + (2 × 9) = 8 + 18 = 26___

Perimeter: ___26___ units

I can write expressions to represent the side lengths for rectangles with an area of 36 square units.

The expressions are 1 × 36, 2 × 18, 3 × 12, 4 × 9, and 6 × 6.

For the first rectangle, I choose to draw a rectangle with side lengths of 3 units and 12 units.

I multiply the width times the length to find the area. The area is 36 square units.

Opposite side lengths of rectangles are equal, so I multiply the width by 2 and the length by 2. Then I add the products to find the perimeter. The perimeter is 30 units.

For the second rectangle, I choose to draw a rectangle with side lengths of 4 units and 9 units.

The area is 36 square units and the perimeter is 26 units.

Copyright © Great Minds PBC

4. The perimeter of a square picture frame is 20 inches. What is the length of 1 side of the picture frame?

5 inches

I know that a square has 4 equal sides. The unknown is 1 side length.

? inches

All 4 equal sides total 20 inches.

I can write an equation to determine the length of 1 side.

$$20 \div 4 = 5$$

Each side has a length of 5 inches.

Copyright © Great Minds PBC

Name _____ Date _____

Add. Show your thinking.

1. $4\frac{2}{8} + \frac{4}{8} =$ _____

2. $3\frac{5}{12} + \frac{7}{12} =$ _____

3. $\frac{4}{6} + 2\frac{5}{6} =$ _____

4. $\frac{7}{8} + 1\frac{4}{8} =$ _____

REMEMBER

5. Draw and shade 2 rectangles that both have an area of 24 square units but have different perimeters. Show how to find the area and perimeter of each rectangle.

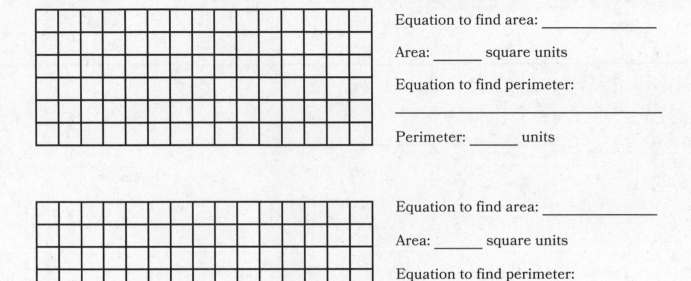

Equation to find area: _____

Area: _____ square units

Equation to find perimeter:

_____

Perimeter: _____ units

Equation to find area: _____

Area: _____ square units

Equation to find perimeter:

_____

Perimeter: _____ units

6. The perimeter of a square window is 8 feet. What is the length of 1 side of the window?

Copyright © Great Minds PBC

24

Name _____   Date _____

Add. Show your thinking.

1. $4\frac{3}{6} + 3\frac{3}{6} = \underline{\quad 8 \quad}$

$$4\frac{3}{6} \xrightarrow{\;+\frac{3}{6}\;} 5 \xrightarrow{\;+3\;} 8$$

> I can count on by using benchmark numbers. I decompose $3\frac{3}{6}$ into 3 and $\frac{3}{6}$.
>
> I can use the arrow way to show my thinking.
>
> I add $\frac{3}{6}$ to $4\frac{3}{6}$ to make the next whole number, 5.
>
> $$4\frac{3}{6} \xrightarrow{\;+\frac{3}{6}\;} 5$$
>
> Then I add 3 to 5.
>
> $$5 \xrightarrow{\;+3\;} 8$$

2. $2\frac{7}{12} + 6\frac{11}{12} = \underline{\quad 9\frac{6}{12} \quad}$

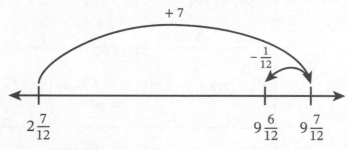

> I can use compensation and a number line to help me solve the problem.
>
> I know $6\frac{11}{12}$ is close to 7, so I add 7 to $2\frac{7}{12}$ to get $9\frac{7}{12}$.
>
> But now I have added $\frac{1}{12}$ too much, so I subtract $\frac{1}{12}$ from $9\frac{7}{12}$.
>
> So $2\frac{7}{12} + 6\frac{11}{12} = 9\frac{6}{12}$.

Copyright © Great Minds PBC

REMEMBER

3. Multiply. Show your method.

5,271 × 3 = __15,813__

	5,000	200	70	1
3	15,000	600	210	3

15,000 + 600 + 210 + 3 = 15,813

I can draw an area model to help me multiply.

I break apart 5,271 into place value units. I break apart the area of the rectangle into parts for each unit. I write the units at the top of each part.

I label the side 3 since I am multiplying by 3.

I multiply each part by 3 to find the partial products.

	5,000	200	70	1
3	5,000 × 3 = 15,000	200 × 3 = 600	70 × 3 = 210	1 × 3 = 3

I add the partial products.

The sum of the partial products is equal to the area of the entire rectangle.

Copyright © Great Minds PBC

4. Divide. Show your method.

$128 ÷ 4 = \underline{32}$

I can decompose 128 into 120 and 8 to use facts I know to help me divide.

$$128 ÷ 4 = (120 + 8) ÷ 4$$
$$= (120 ÷ 4) + (8 ÷ 4)$$
$$= 30 + 2$$
$$= 32$$

24

Name _____   Date _____

Add. Show your thinking.

1. $3\frac{4}{10} + 4\frac{5}{10} = $ _____

2. $2\frac{2}{3} + 5\frac{1}{3} = $ _____

3. $1\frac{3}{5} + 3\frac{4}{5} = $ _____

Copyright © Great Minds PBC

REMEMBER

4. Multiply. Show your method.

$4{,}358 \times 2 = \underline{\hspace{1.5cm}}$

5. Divide. Show your method.

$156 \div 3 = \underline{\hspace{1.5cm}}$

Copyright © Great Minds PBC

💬💬 **25**

Name _____ Date _____

1. Subtract by counting on. Use the arrow way or a number line to help you. Then complete the equations.

$$2\frac{1}{5} - \frac{3}{5} = \underline{\quad 1\frac{3}{5} \quad}$$

$$\frac{3}{5} + \underline{\quad 1\frac{3}{5} \quad} = 2\frac{1}{5}$$

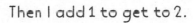

$$\frac{3}{5} \xrightarrow{+\frac{2}{5}} 1 \xrightarrow{+1} 2 \xrightarrow{+\frac{1}{5}} 2\frac{1}{5}$$

> I can use the arrow way to help me count on to find the difference.
>
> I add $\frac{2}{5}$ to $\frac{3}{5}$ to get to 1.
>
> $$\frac{3}{5} \xrightarrow{+\frac{2}{5}} 1$$
>
> Then I add 1 to get to 2.
>
> $$\frac{3}{5} \xrightarrow{+\frac{2}{5}} 1 \xrightarrow{+1} 2$$
>
> Next, I add $\frac{1}{5}$ to 2 to make the total, $2\frac{1}{5}$.
>
> $$\frac{3}{5} \xrightarrow{+\frac{2}{5}} 1 \xrightarrow{+1} 2 \xrightarrow{+\frac{1}{5}} 2\frac{1}{5}$$
>
> Last, I add the parts I counted on. So $\frac{2}{5} + 1 + \frac{1}{5} = 1\frac{3}{5}$.

2. Subtract by using compensation. Use the arrow way or a number line to help you. Then complete the equation.

$1\frac{4}{12} - \frac{10}{12} = \underline{\frac{6}{12}}$

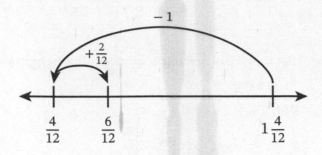

$\frac{4}{12}$ $\frac{6}{12}$ $1\frac{4}{12}$

I can use a number line to show compensation to find the difference. Because $\frac{10}{12}$ is close to 1, I subtract 1 from $\frac{14}{12}$.

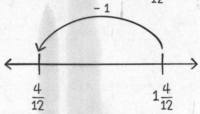

$\frac{4}{12}$ $1\frac{4}{12}$

I add back $\frac{2}{12}$ so that I only subtract $\frac{10}{12}$.

So $1\frac{4}{12} - \frac{10}{12} = \frac{6}{12}$.

3. Subtract. Show your thinking.

$2\frac{2}{6} - \frac{5}{6} = \underline{1\frac{3}{6}}$

$2\frac{2}{6} - \frac{5}{6}$

$\frac{12}{6} \quad \frac{2}{6}$

$\frac{12}{6} + \frac{2}{6} = \frac{14}{6}$

$\frac{14}{6} - \frac{5}{6} = \frac{9}{6}$

$1\frac{3}{6} = \frac{9}{6}$

I can draw a number bond to help me rename the mixed number, $2\frac{2}{6}$, as a fraction greater than 1.

$2\frac{2}{6}$

$\frac{12}{6} \quad \frac{2}{6}$

$2\frac{2}{6} = \frac{14}{6}$

Now I can subtract.

$\frac{14}{6} - \frac{5}{6} = \frac{9}{6}$

I rename the fraction greater than 1 as a mixed number.

$\frac{9}{6} = 1\frac{3}{6}$

Copyright © Great Minds PBC

REMEMBER

4. Complete the conversion table.

Pints	Cups
1	2
5	10
7	14
16	32

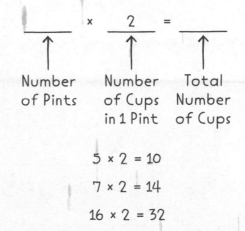

I know that 1 pint is 2 times as much as 1 cup.

1 pint = 2 cups

I can convert pints to cups by multiplying the number of pints by 2.

_____ × _2_ = _____

↑ Number of Pints   ↑ Number of Cups in 1 Pint   ↑ Total Number of Cups

5 × 2 = 10

7 × 2 = 14

16 × 2 = 32

5. Amy has 6 gallons 2 quarts of lemonade in pitchers. She serves 17 quarts of lemonade at a banquet.

How many quarts of lemonade remain in the pitchers?

$$6 \times 4 = 24$$
$$24 + 2 = 26$$
$$26 - 17 = 9$$

There are 9 quarts of lemonade remaining in the pitchers.

First, I convert the gallons to quarts.

I know there are 4 quarts in 1 gallon.

I can multiply the number of gallons by 4 to find the total number of quarts.

6 × 4 = 24

Next, I add 24 and 2 to find how many total quarts Amy has before the banquet.

24 + 2 = 26

She serves 17 quarts of lemonade. I subtract the quarts she serves from the total quarts to find how much lemonade remains in the pitchers.

Copyright © Great Minds PBC

Name _____ Date _____

1. Subtract by counting on. Use the arrow way or a number line to help you. Then complete the equations.

   $3\frac{2}{7} - \frac{5}{7} = $ _____

   $\frac{5}{7} + $ _____ $= 3\frac{2}{7}$

2. Subtract by using compensation. Use the arrow way or a number line to help you. Then complete the equation.

   $2\frac{4}{9} - \frac{8}{9} = $ _____

Subtract. Show your thinking.

3. $1\frac{3}{8} - \frac{6}{8} = $ _____

4. $2\frac{2}{6} - \frac{5}{6} = $ _____

REMEMBER

5. Complete the conversion table.

Pints	Cups
1	2
3	
6	
15	

6. Carla has 13 gallons 1 quart of gas in her gas storage tank. She uses 38 quarts of gas to fill the tank of her snowmobile. How many quarts of gas remain in the storage tank?

Copyright © Great Minds PBC

Name _____ Date _____

1. Decompose the part to subtract. Use the arrow way to help you. Then complete the equation.

$2\frac{3}{12} - \frac{9}{12} = \underline{1\frac{6}{12}}$

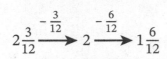

$2\frac{3}{12} \xrightarrow{-\frac{3}{12}} 2 \xrightarrow{-\frac{6}{12}} 1\frac{6}{12}$

I can decompose the part, $\frac{9}{12}$, to help me subtract.

I break apart $\frac{9}{12}$ into $\frac{3}{12}$ and $\frac{6}{12}$ to subtract the $\frac{3}{12}$ first.

$$2\frac{3}{12} - \frac{9}{12}$$

$$\frac{3}{12} \quad \frac{6}{12}$$

Then I subtract $\frac{6}{12}$ from 2. I use the arrow way to show my thinking.

2. Decompose the total to subtract from 1. Then complete the equation.

$1\frac{2}{10} - \frac{6}{10} = \underline{\frac{6}{10}}$

I decompose the total, $1\frac{2}{10}$, to subtract $\frac{6}{10}$ from 1.

I break apart $1\frac{2}{10}$ into $\frac{2}{10}$ and 1. I subtract $\frac{6}{10}$ from 1. I know that $1 = \frac{10}{10}$, so $\frac{10}{10} - \frac{6}{10} = \frac{4}{10}$.

$$1\frac{2}{10} - \frac{6}{10}$$

$$\frac{2}{10} \quad 1 \quad \frac{4}{10}$$

I add the remaining parts.

$$\frac{2}{10} + \frac{4}{10} = \frac{6}{10}$$

Copyright © Great Minds PBC

3. Rename the total to subtract. Then complete the equation.

$2\frac{1}{10} - \frac{3}{10} = \underline{1\frac{8}{10}}$

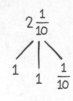

I can decompose $2\frac{1}{10}$ into 1, 1, and $\frac{1}{10}$.

I know that $1 = \frac{10}{10}$, so $2\frac{1}{10}$ is the same as $1 + \frac{10}{10} + \frac{1}{10}$. I add the fractions to rename $2\frac{1}{10}$ as 1 and $\frac{11}{10}$.

Now I subtract: $\frac{11}{10} - \frac{3}{10} = \frac{8}{10}$.

I add the remaining parts.

$$1 + \frac{8}{10} = 1\frac{8}{10}$$

REMEMBER

4. James has 10 yards of fabric. The fabric is cut into 5 equal pieces. What is the length, in feet, of each piece of fabric?

The length of each piece of fabric is 6 feet.

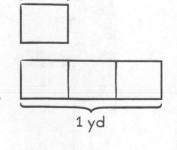

I know 1 yard = 3 feet.

I can multiply 10 yards by 3 to find the total number of feet.

$$10 \times 3 = 30$$

The fabric is 30 feet long.

I can divide 30 by 5 to find the number of feet in each piece of fabric.

$$30 \div 5 = 6$$

The length of piece of fabric is 6 feet.

1 ft

1 yd

                          Copyright © Great Minds PBC

/ 26

Name _____ Date _____

1. Decompose the part to subtract. Use the arrow way to help you. Then complete the equation.

$2\frac{2}{5} - \frac{3}{5} =$ _____

2. Decompose the total to subtract from 1. Then complete the equation.

$1\frac{3}{8} - \frac{6}{8} =$ _____

3. Rename the total to subtract. Then complete the equation.

$2\frac{1}{10} - \frac{4}{10} =$ _____

REMEMBER

4. Casey has 16 yards of fencing. It is in 4 equal sections. What is the length, in feet, of each section of fencing?

Copyright © Great Minds PBC

27

Name _____ Date _____

1. Subtract by counting on. Use the arrow way or a number line to help you. Then complete the equations.

$$6\frac{2}{10} - 1\frac{7}{10} = \underline{\quad 4\frac{5}{10} \quad}$$

$$1\frac{7}{10} + \underline{\quad 4\frac{5}{10} \quad} = 6\frac{2}{10}$$

I can think of subtraction as addition with an unknown addend.

$$1\frac{7}{10} + \underline{\qquad} = 6\frac{2}{10}$$

I can use a number line to count on. I start at $1\frac{7}{10}$. I can add $\frac{3}{10}$ to make 2. Then I add $4\frac{2}{10}$ to make $6\frac{2}{10}$.

I add the amounts I counted on.

$$\frac{3}{10} + 4\frac{2}{10} = 4\frac{5}{10}$$

I can also show my work by using the arrow way.

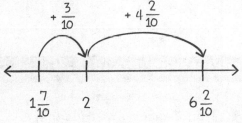

$$1\frac{7}{10} \xrightarrow{+\frac{3}{10}} 2 \xrightarrow{+4\frac{2}{10}} 6\frac{2}{10}$$

2. Subtract by using compensation. Use the arrow way or a number line to help you. Then complete the equation.

$$8\frac{3}{8} - 1\frac{7}{8} = \underline{\quad 6\frac{4}{8} \quad}$$

I can subtract by using compensation and a number line.

$1\frac{7}{8}$ is close to 2. I know that $8\frac{3}{8} - 2 = 6\frac{3}{8}$.

I subtracted $\frac{1}{8}$ more than I needed, so I'll add back $\frac{1}{8}$.

The difference is $6\frac{4}{8}$.

I can also show my work by using the arrow way.

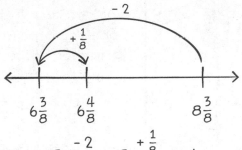

$$8\frac{3}{8} \xrightarrow{-2} 6\frac{3}{8} \xrightarrow{+\frac{1}{8}} 6\frac{4}{8}$$

Copyright © Great Minds PBC

3. Decompose the total to subtract from a whole number. Then complete the equation.

$8\frac{1}{4} - 1\frac{3}{4} = \underline{6\frac{2}{4}}$

I decompose the total to subtract from a whole number.

$1\frac{3}{4}$ is close to 2. I break apart $8\frac{1}{4}$ into $6\frac{1}{4}$ and 2.

I know that $2 - 1\frac{3}{4} = \frac{1}{4}$.

I add the remaining parts.

$6\frac{1}{4} + \frac{1}{4} = 6\frac{2}{4}$

REMEMBER

4. Ivan measures a string that is 5 times as long as Gabe's string. Ivan's string is 35 inches long. How long is Gabe's string?

$35 \div 5 = 7$

Gabe's string is 7 inches long.

I read the problem. I read again.

As I reread, I think about what I can draw.

I draw a tape diagram with 1 unit to represent the length of Gabe's string. I label it with a question mark because it is the unknown.

I draw 5 units to represent that Ivan's string is 5 times as long as Gabe's string. Ivan's string is 35 inches long, so I label the 5 units with a total of 35.

I need to find the length of Gabe's string.

I can divide 35 by 5 to find the length of Gabe's string.

Copyright © Great Minds PBC

Name _____ Date _____

1. Subtract by counting on. Use the arrow way or a number line to help you. Then complete the equations.

$7\frac{4}{6} - 2\frac{5}{6} =$ _____

$2\frac{5}{6} +$ _____ $= 7\frac{4}{6}$

2. Subtract by using compensation. Use the arrow way or a number line to help you. Then complete the equation.

$8\frac{4}{10} - 1\frac{9}{10} =$ _____

Decompose the total to subtract from a whole number. Then complete the equation.

3. $6\frac{1}{5} - 1\frac{2}{5} =$ _____

Copyright © Great Minds PBC

REMEMBER

Use the Read–Draw–Write process to solve the problem.

4. Pablo reads 3 times as many pages as David. Pablo reads 21 pages. How many pages does David read?

Copyright © Great Minds PBC

28

Name _____ Date _____

Use the Read–Draw–Write process to solve each problem.

1. Ray swims $1\frac{6}{10}$ kilometers on Monday and $2\frac{5}{10}$ kilometers on Tuesday. How many kilometers does he swim in total?

$$1\frac{6}{10} + 2\frac{5}{10} = 4\frac{1}{10}$$

Ray swims $4\frac{1}{10}$ kilometers in total.

I read the problem. I read again.

As I reread, I think about what I can draw.

I draw a tape diagram to represent the $1\frac{6}{10}$ kilometers Ray swims on Monday and the $2\frac{5}{10}$ kilometers he swims on Tuesday.

The unknown is how many kilometers he swims in total.

I can add the 2 parts to find how many kilometers Ray swims in total.

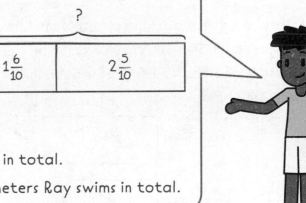

2. Deepa drinks $3\frac{3}{5}$ liters of water in the morning and $4\frac{2}{5}$ liters of water in the afternoon.

   a. How many more liters of water does she drink in the afternoon than in the morning?

   $$4\frac{2}{5} - 3\frac{3}{5} = \frac{4}{5}$$

   She drinks $\frac{4}{5}$ liters more water in the afternoon.

   b. How many liters of water does she drink in total?

   $$4\frac{2}{5} + 3\frac{3}{5} = 7\frac{5}{5} = 8$$

   She drinks 8 liters of water in total.

> I read the problem. I read again.
>
> As I reread, I think about what I can draw.
>
> I draw a tape diagram to represent the $3\frac{3}{5}$ liters of water Deepa drinks in the morning and the $4\frac{2}{5}$ liters of water she drinks in the afternoon.
>
> Morning $\boxed{4\frac{2}{5}}$
>
> Afternoon $\boxed{3\frac{3}{5}}$

> For part (a), the unknown is the difference between the amount of water she drank in the morning and the amount she drank in the afternoon.
>
> Morning $\boxed{4\frac{2}{5}}$
>
> Afternoon $\boxed{3\frac{3}{5}}$?
>
> I can subtract to find the unknown.

> For part (b), the unknown is the amount of liters she drinks in total.
>
> Morning $\boxed{4\frac{2}{5}}$
>
> Afternoon $\boxed{3\frac{3}{5}}$ ⎱ ?
>
> I can add to find the unknown.

Copyright © Great Minds PBC

REMEMBER

3. Compare the fractions by using >, =, or <.

    Explain your thinking by using pictures, numbers, or words.

    $\frac{1}{2}$ ___<___ $\frac{4}{5}$

    $$\frac{1}{2} = \frac{5 \times 1}{5 \times 2} = \frac{5}{10}$$

    $$\frac{4}{5} = \frac{2 \times 4}{2 \times 5} = \frac{8}{10}$$

    $\frac{5}{10} < \frac{8}{10}$, so $\frac{1}{2} < \frac{4}{5}$.

I can find a common denominator to compare fractions.

I know that 10 is a multiple of 2 and 5.

I rename both fractions as tenths.

To compare the numerators, I can look at the number of units.

I know $\frac{5}{10}$ is less than $\frac{8}{10}$ because 5 is less than 8.

So $\frac{1}{2} < \frac{4}{5}$.

Name

Date

Use the Read–Draw–Write process to solve each problem.

1. Ray practices piano for $2\frac{1}{2}$ hours on Saturday and $3\frac{1}{2}$ hours on Sunday. How many total hours does he practice piano?

2. Amy runs $4\frac{3}{4}$ miles on Saturday and $6\frac{1}{4}$ miles on Sunday.

   a. How many more miles does she run on Sunday than on Saturday?

   b. How many miles does she run in total?

Copyright © Great Minds PBC

REMEMBER

3. Compare the fractions by using >, =, or <. Explain your thinking by using pictures, numbers, or words.

$$\frac{3}{4} \underline{\hspace{1cm}} \frac{2}{6}$$

Copyright © Great Minds PBC

Name _____   Date _____

29

1. Shen squeezes the juice from some oranges. He measures how much juice came from each orange. He makes a line plot from the data.

Amount of Juice in Shen's Oranges

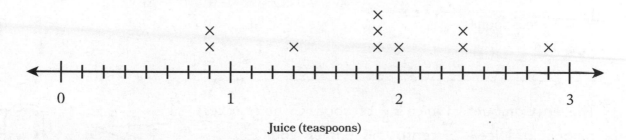

Juice (teaspoons)

a. How many oranges does Shen measure the juice from?

   10 oranges

b. What is the greatest amount of juice from 1 of Shen's oranges?

   $2\frac{7}{8}$ teaspoons

> I count the number of Xs to find the number of oranges Shen measures the juice from. There are 10 Xs, so I know that Shen measures the juice from 10 oranges.
>
> The greatest amount is the number closest to 3 on the number line. I see an X above $2\frac{7}{8}$. The greatest amount of juice is $2\frac{7}{8}$ teaspoons.

Copyright © Great Minds PBC

c. Use >, =, or < to compare the least amount of juice to the greatest amount of juice.

$\frac{7}{8} < 2\frac{7}{8}$

d. How many oranges have more than 2 teaspoons of juice?

3 oranges

e. What is the total amount of juice from the 2 oranges with the least amount of juice?

$$\frac{7}{8} + \frac{7}{8} = \frac{14}{8}$$

$\frac{14}{8}$ or $1\frac{6}{8}$ teaspoons

The least amount of juice is $\frac{7}{8}$ teaspoons. The greatest amount of juice is $2\frac{7}{8}$ teaspoons. $\frac{7}{8}$ is less than $2\frac{7}{8}$.

To find how many oranges gave more than 2 teaspoons of juice, I count the number of Xs located to the right of 2 on the line plot. There are 3 Xs, so 3 oranges have more than 2 teaspoons of juice.

The 2 oranges with the least amount of juice each have $\frac{7}{8}$ teaspoons.

$$\frac{7}{8} + \frac{7}{8} = \frac{14}{8} = 1\frac{6}{8}$$

                          Copyright © Great Minds PBC

REMEMBER

Use the Read–Draw–Write process to solve the problem.

2. Box A has 72 paper clips in it. Box A has 4 times as many paper clips as box B. How many paper clips does box B have?

$72 \div 4 = 18$

Box B has 18 paper clips.

I read the problem. I read again.

As I reread, I think about what I can draw.

I draw a tape diagram with 2 tapes. There is 1 tape for box A and 1 tape for box B.

I draw the tape for box A to show box A has 4 times as many paper clips as box B. I label the tape for box A to show the total of 72 paper clips. I need to find the number of paper clips in box B, so I label it with a question mark.

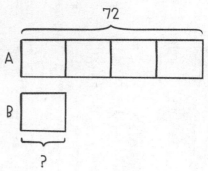

I can divide the total number of paper clips in box A by 4 to find the number of paper clips in box B.

3. Gabe draws a quadrilateral that has 2 pairs of parallel sides, 4 equal side lengths, and 4 right angles. Which shape could Gabe have drawn? Circle all the shapes that could name Gabe's quadrilateral.

    A. Square

    B. Rectangle

    C. Trapezoid

    D. Parallelogram

    E. Rhombus

I know each shape listed is a quadrilateral. I can think about other attributes of each shape.

square	rectangle	trapezoid	parallelogram	rhombus
• 2 pairs of parallel sides	• 2 pairs of parallel sides	• At least 1 pair of parallel sides	• 2 pairs of parallel sides	• 2 pairs of parallel sides
• 4 sides of equal length	• Opposite side lengths are equal			• 4 sides of equal length
• 4 right angles	• 4 right angles			

Gabe's quadrilateral has 2 pairs of parallel sides. A square, rectangle, parallelogram, and a rhombus all have 2 pairs of parallel sides. A trapezoid has at least 1 pair of parallel sides, which means it can have 1 or more pairs of parallel sides. That information does not help me eliminate any of the shapes.

Gabe's quadrilateral has 4 equal sides. A trapezoid, parallelogram, rhombus, and square all can have 4 equal sides. The opposite side lengths in a rectangle are equal, so it could have 4 equal sides. That information does not help me eliminate any shapes.

Gabe's quadrilateral also has 4 right angles. All of the shapes could have 4 right angles. I cannot eliminate any shapes.

That means that a square, a rectangle, a trapezoid, a parallelogram and a rhombus can all name Gabe's quadrilateral.

Copyright © Great Minds PBC

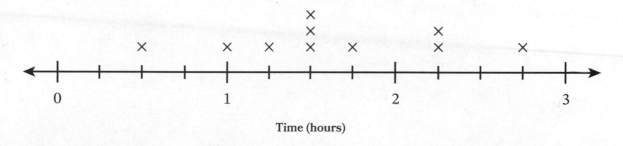

Name _____ Date _____

1. The students in Mr. Endo's class measure the time they spend walking their dogs in 1 week. They make a line plot from the data.

Time Mr. Endo's Students Spend Walking Their Dogs

Time (hours)

a. How many students in Mr. Endo's class walk their dog?

b. What is the shortest amount of time a student walks their dog?

c. Use >, =, or < to compare the shortest amount of time to the longest amount of time a student walks their dog.

d. How many students walk their dogs for at least 2 hours?

e. What is the total amount of time for all the students who walk a dog for $2\frac{1}{4}$ hours?

REMEMBER

Use the Read–Draw–Write process to solve the problem.

2. Mrs. Smith has 96 bottles of paint. She has 8 times as many bottles of paint as paintbrushes. How many paintbrushes does Mrs. Smith have?

3. Carla draws a quadrilateral that has 2 pairs of parallel sides and all sides do not have the same length. Which shape could Carla have drawn? Circle all the shapes that could name Carla's quadrilateral.

   A. Square

   B. Rectangle

   C. Trapezoid

   D. Triangle

   E. Rhombus

Copyright © Great Minds PBC

Name _____ Date _____

1. The finishing times for a bicycle race are shown in the table.

   a. Use the data in the table to make a line plot.

Finishing Times of Racers

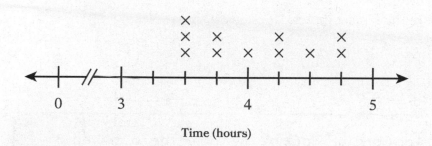

Time (hours)

Racer	Time (hours)
Liz	$3\frac{3}{4}$
Gabe	$3\frac{2}{4}$
Mia	4
Ray	$4\frac{1}{4}$
Pablo	$3\frac{2}{4}$
Luke	$4\frac{2}{4}$
Deepa	$4\frac{3}{4}$
Carla	$4\frac{1}{4}$
Casey	$3\frac{3}{4}$
Jayla	$4\frac{3}{4}$
Ivan	$3\frac{2}{4}$

I start by creating a scale for the line plot. The fastest time is $3\frac{2}{4}$ hours, and the slowest time is $4\frac{3}{4}$ hours. I use the whole numbers 3 and 5 to start and end the scale of my line plot. I label the whole number 4 as well. Every mixed number in the table has fourths as the fractional unit. So I partition each whole number interval into fourths.

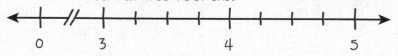

I make the line plot by representing each time in the table on the line plot with an X. The scale represents the time in hours, so I write **Time (hours)** below the line plot.

I write a title for the line plot to show that the times represent the finishing times of the racers.

Copyright © Great Minds PBC

b. Which racer finished the race in 4 hours?

Mia

c. What is the most frequent time it took to finish the race?

$3\frac{2}{4}$ hours

d. How many more racers finished in $3\frac{2}{4}$ hours than in $4\frac{2}{4}$ hours?

2 racers

e. Use >, =, or < to compare the fastest time to the slowest time.

$3\frac{2}{4} < 4\frac{3}{4}$

For part (b), I look at the table. The racer who finished in 4 hours is Mia.

For part (c), I look at the line plot. The time with the most Xs above it is $3\frac{2}{4}$ hours.

For part (d), I look at the line plot to see how many racers finished in $3\frac{2}{4}$ hours. I see 3 Xs above $3\frac{2}{4}$ hours. I look at the line plot again to see how many racers finished in $4\frac{2}{4}$ hours. I see 1 X above $4\frac{2}{4}$ hours.

Because 3 − 1 = 2, there are 2 more racers who finished in $3\frac{2}{4}$ hours than in $4\frac{2}{4}$ hours.

For part (e), I look at the smallest and largest values that have an X above them. The smallest value, $3\frac{2}{4}$, represents the fastest time. The largest value, $4\frac{3}{4}$, represents the slowest time. $3\frac{2}{4}$ is less than $4\frac{3}{4}$.

                           Copyright © Great Minds PBC

REMEMBER

2. Complete the table.

Yards	Feet
2	6
3	9
5	15
7	21

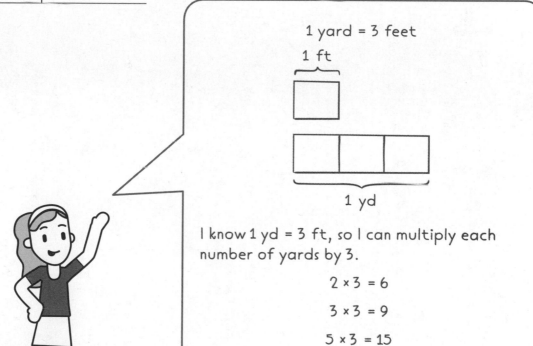

1 yard = 3 feet

1 ft

1 yd

I know 1 yd = 3 ft, so I can multiply each number of yards by 3.

$$2 \times 3 = 6$$

$$3 \times 3 = 9$$

$$5 \times 3 = 15$$

$$7 \times 3 = 21$$

Copyright © Great Minds PBC

Name _____ Date _____

1. Eva is a scientist studying cone snails. She records the lengths of cone snail shells in the table.

Cone Snail	A	B	C	D	E	F	G	H	I	J
Shell Length (inches)	$1\frac{1}{8}$	1	$\frac{7}{8}$	$1\frac{1}{8}$	$1\frac{2}{8}$	$\frac{7}{8}$	$\frac{3}{8}$	$1\frac{1}{8}$	$\frac{3}{8}$	$\frac{6}{8}$

a. Use the data in the table to make a line plot.

b. Which cone snail has a shell length of 1 inch?

c. What is the most frequent shell length?

d. What is the difference between the longest and shortest shell lengths?

e. Use >, =, or < to compare the shell lengths of cone snails C and I.

REMEMBER

2. Complete the table.

Yards	Feet
4	
8	
9	
12	

Copyright © Great Minds PBC

FAMILY MATH
Repeated Addition of Fractions as Multiplication

Dear Family,

Your student is using familiar models to multiply fractions and mixed numbers by a whole number. Students break apart a fraction to write repeated addition equations and a multiplication problem. They use methods that were used earlier in the year with whole numbers to break apart multiplication problems in different ways. Students solve word problems and use the context of the problem to decide when to rename a product, that is a fraction greater than 1, as a mixed number.

$$\frac{7}{8} = 7 \times \frac{1}{8}$$

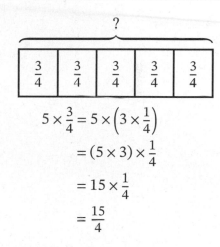

$$\frac{7}{8} = \frac{1}{8} + \frac{1}{8} + \frac{1}{8} + \frac{1}{8} + \frac{1}{8} + \frac{1}{8} + \frac{1}{8}$$

$$4 \times \frac{2}{5} = 4 \times \left(2 \times \frac{1}{5}\right)$$
$$= (4 \times 2) \times \frac{1}{5}$$
$$= 8 \times \frac{1}{5}$$
$$= \frac{8}{5}$$

$$5 \times \frac{3}{4} = 5 \times \left(3 \times \frac{1}{4}\right)$$
$$= (5 \times 3) \times \frac{1}{4}$$
$$= 15 \times \frac{1}{4}$$
$$= \frac{15}{4}$$

$\frac{7}{8}$ is equal to adding $\frac{1}{8}$, 7 times. Similar to whole numbers, repeated addition can be written by using multiplication.

Grouping numbers differently may help students multiply. In this case, they multiply the whole numbers first and then the unit fraction.

Students use tape diagrams to make sense of word problems. They write and solve multiplication equations that describe their tape diagram to answer the question in a word problem.

At-Home Activity

Mixed Number Multiplication

Help your student use a favorite recipe to practice multiplying a mixed number and a whole number. Invite your student to research a recipe for a favorite dish and record an ingredient amount that is a mixed number from the recipe. Then have your student multiply the mixed number by 2 to see how much of the ingredient would be needed to double the recipe. Repeat this process by multiplying the amount by other whole numbers to see how much of the ingredient would be needed to triple or quadruple the recipe.

Copyright © Great Minds PBC

Name _____   Date _____

1. Complete the equation and statement. Then represent the equation with a tape diagram.

$$\frac{7}{8} = \frac{1}{8} + \frac{1}{8} + \frac{1}{8} + \frac{1}{8} + \frac{1}{8} + \frac{1}{8} + \frac{1}{8}$$

$$\frac{7}{8} = \underline{\quad 7 \quad} \times \boxed{\frac{1}{8}}$$

$\frac{7}{8}$ is the ___seventh___ multiple of ___$\frac{1}{8}$___ .

$\boxed{\frac{7}{8}}$

$\frac{1}{8}$	$\frac{1}{8}$	$\frac{1}{8}$	$\frac{1}{8}$	$\frac{1}{8}$	$\frac{1}{8}$	$\frac{1}{8}$

I can decompose $\frac{7}{8}$ to represent it as the product of a whole number and a unit fraction.

$\frac{7}{8}$ can be represented by the multiplication equation $\frac{7}{8} = 7 \times \frac{1}{8}$.

If I count by $\frac{1}{8}$ seven times then I'll get $\frac{7}{8}$. So I can describe $\frac{7}{8}$ as the seventh multiple of $\frac{1}{8}$.

I can represent the decomposition with a tape diagram. I draw 7 parts and each part has a value of $\frac{1}{8}$. The total of $\frac{7}{8}$ is shown in the tape diagram is.

2. Complete the equation and statement. Then represent the equation with a number line.

$\frac{11}{8} = \frac{1}{8} + \frac{1}{8} + \frac{1}{8} + \frac{1}{8} + \frac{1}{8} + \frac{1}{8} + \frac{1}{8} + \frac{1}{8} + \frac{1}{8} + \frac{1}{8} + \frac{1}{8}$

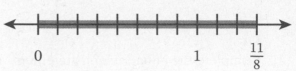

$\frac{11}{8} = \underline{\quad 11 \quad} \times \dfrac{1}{8}$

$\frac{11}{8}$ is the __eleventh__ multiple of $\underline{\dfrac{1}{8}}$.

I can decompose $\frac{11}{8}$ to represent the fraction as the product of a whole number and a unit fraction.

$\frac{11}{8}$ can be represented by the multiplication equation $\frac{11}{8} = 11 \times \frac{1}{8}$.

If I count by $\frac{1}{8}$ eleven times then I'll get $\frac{11}{8}$. So I can describe $\frac{11}{8}$ as the eleventh multiple of $\frac{1}{8}$.

I can represent the decomposition with a number line. The number line starts at 0 and ends at $\frac{11}{8}$. I draw 11 units and each unit is $\frac{1}{8}$.

REMEMBER

3. Fill in the blanks to complete the number line and the statement.

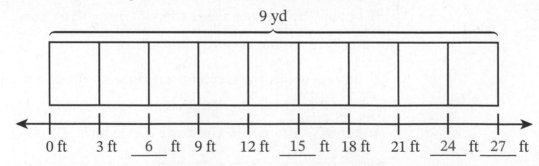

9 yd

0 ft   3 ft   _6_ ft   9 ft   12 ft   _15_ ft   18 ft   21 ft   _24_ ft _27_ ft

There are __27__ feet in 9 yards.

A yard is 3 times as long as a foot.

Each interval on the number line represents 3 feet. I start at 3 feet and skip-count by threes.

The last tick mark is 27 feet.

I see that there are 27 feet in 9 yards.

                                    Copyright © Great Minds PBC

4. The length of a rectangular mirror is 14 inches. The perimeter is 38 inches.

What is the width?

The width of the mirror is 5 inches.

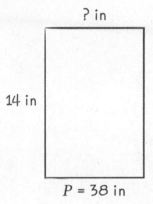

I know the length of 1 side of the mirror and its perimeter. I need to find the width of the mirror.

A formula for the perimeter of a rectangle is $P = 2(l + w)$.

I use the formula to find the width.

$$38 = 2 \times (14 + w)$$

I know the perimeter is the sum of twice the length and twice the width.

First, I find half of the perimeter. Half of 38 is 19.

$$19 = 14 + w$$

The width is 5 inches.

$$19 = 14 + 5$$

Copyright © Great Minds PBC

Name _____   Date _____

1. Complete the equation and statement. Then represent the equation with a tape diagram.

$$\frac{6}{8} = \frac{1}{8} + \frac{1}{8} + \frac{1}{8} + \frac{1}{8} + \frac{1}{8} + \frac{1}{8}$$

$$\frac{6}{8} = \underline{\hspace{1cm}} \times \frac{\blacksquare}{8}$$

$\frac{6}{8}$ is the _____ multiple of _____.

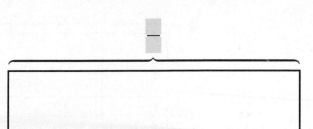

2. Complete the equation and statement. Then represent the equation with a number line.

$$\frac{8}{5} = \frac{1}{5} + \frac{1}{5} + \frac{1}{5} + \frac{1}{5} + \frac{1}{5} + \frac{1}{5} + \frac{1}{5} + \frac{1}{5}$$

$$\frac{8}{5} = \underline{\hspace{1cm}} \times \frac{\blacksquare}{\blacksquare}$$

$\frac{8}{5}$ is the _____ multiple of _____.

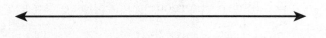

Complete the equations to express each fraction as a whole number times a unit fraction.

3. $\frac{5}{10} = \underline{\hspace{1cm}} \times \frac{\blacksquare}{10}$

4. $\frac{6}{6} = \underline{\hspace{1cm}} \times \frac{\blacksquare}{6}$

5. $\frac{4}{3} = \underline{\hspace{1cm}} \times \frac{\blacksquare}{3}$

REMEMBER

6. Fill in the blanks to complete the number line and the statement.

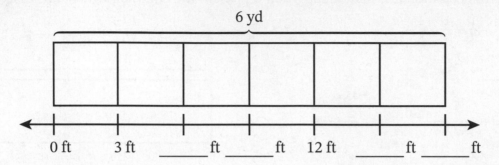

There are _____ feet in 6 yards.

7. The length of a rectangular book cover is 9 inches. The perimeter is 30 inches.

   What is the width?

Copyright © Great Minds PBC

32

Name _____ Date _____

1. Label the number line and complete each statement.

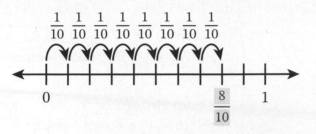

$$\frac{1}{10} \quad \frac{1}{10} \quad \frac{1}{10} \quad \frac{1}{10} \quad \frac{1}{10} \quad \frac{1}{10} \quad \frac{1}{10} \quad \frac{1}{10}$$

0             $\frac{8}{10}$ 1

a. 8×1 tenth $= (\underline{\;\;8\;\;} \times \underline{\;\;1\;\;})$ tenths

              $= \underline{\;\;8\;\;}$ tenths

b. $8 \times \dfrac{1}{10} = \dfrac{8 \times 1}{10}$

           $= \dfrac{8}{10}$

> I count 8 iterations of $\frac{1}{10}$ on the number line. I label $\frac{8}{10}$ on number line.
>
> I use the associative property and parentheses to group the 1 in 1 tenth with the number of groups. So (8×1) tenth = 8 tenths.
>
> I represent the equation as a fraction with a multiplication expression as the numerator, and I complete the equation. So $\frac{8 \times 1}{10} = \frac{8}{10}$.

2. Choose a method and find the product.

$$9 \times \frac{1}{6} = \frac{9 \times 1}{6} = \frac{9}{6}$$

> I can group the whole number with the numerator of the fraction and think of the expression as $(9 \times 1) \times \frac{1}{6}$ or (9×1) sixths.
>
> I write $\frac{9 \times 1}{6}$ and multiply the numbers in the numerator to get $\frac{9}{6}$.

Copyright © Great Minds PBC

REMEMBER

Use the Read–Draw–Write process to solve the problem.

3. A teacher sets out 8 board games on a table. Each board game has 32 pieces. 19 pieces fall on the floor.

   a. About how many pieces are still on the table?

   $8 \times 30 = 240$

   $240 - 20 = 220$

   About 220 pieces are still on the table.

   b. Exactly how many pieces are still on the table?

   $8 \times 32 = 256$

   $256 - 19 = 237$

   Exactly 237 pieces are still on the table.

> 32 is close to 30 and 19 is close to 20.
>
> I find 8 groups of 30 to estimate the total number of pieces.
>
> Then I subtract 20 to find how many pieces are still on the table.

> I read the problem. I read again.
>
> As I reread, I think about what I can draw. I draw a tape diagram.
>
> First, I draw a tape to find the total number of pieces. I draw 8 parts to represent the board games. I label 1 part with 32 pieces. I don't know the total number of pieces.

> I find the exact number of pieces still on the table. First, I multiply and find the total number of pieces.
>
> $$\begin{array}{r} 3\ 2 \\ \times\quad 8 \\ \hline {}^{1}2\ 4\ 6 \\ \hline 2\ 5\ 6 \end{array}$$
>
> Then I subtract the pieces that fell on the floor.
>
> $$\begin{array}{r} {}^{4}2\ {}^{16}5\ 6 \\ -\quad 1\ 9 \\ \hline 2\ 3\ 7 \end{array}$$

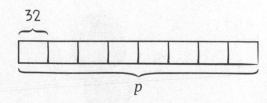

> Next, I draw a tape to show 19 pieces fall. The unknown is how many pieces are left.

c. Is your answer to part (b) reasonable? Explain.

Yes, my estimate of 220 is close to 237, so my answer is reasonable.

I compare my estimate to my answer to see whether it is reasonable. The number of pieces in each game in my estimate is less than the actual number of pieces, so I know my estimate will be less than the exact number.

Name _____   Date _____

Label the number line and complete each statement.

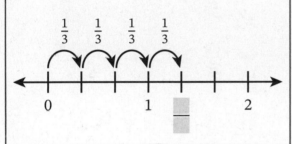

1.

$\frac{1}{6}$ $\frac{1}{6}$ $\frac{1}{6}$ $\frac{1}{6}$ $\frac{1}{6}$

0   ▢/▢   1

a. 5×1 sixth $= (5 \times$ _____ $)$ sixth

= _____ sixths

b. $5 \times \frac{1}{6} = \frac{5 \times ▢}{6}$

$= \frac{▢}{6}$

2.

$\frac{1}{3}$ $\frac{1}{3}$ $\frac{1}{3}$ $\frac{1}{3}$

0   1 ▢/▢   2

a. 4×1 third $= ($ _____ $\times 1)$ third

= _____ thirds

b. $4 \times \frac{1}{3} = \frac{▢ \times ▢}{▢}$

$= \frac{▢}{▢}$

Choose a method and find the product.

3. $13 \times \frac{1}{10}$

4. $15 \times \frac{1}{8}$

REMEMBER

Use the Read–Draw–Write process to solve the problem.

5. An art studio has 47 tables. There are 9 clean paintbrushes on each table. After some artists finish painting, 68 paintbrushes are dirty.

a. About how many paintbrushes are clean?

b. Exactly how many paintbrushes are clean?

c. Is your answer to part (b) reasonable? Explain.

Copyright © Great Minds PBC

Name _____ Date _____

Use the Read–Draw–Write process to solve each problem.

1. A bag of popcorn has a mass of $\frac{3}{10}$ kilograms. A bag of pears has 6 times as much mass as the bag of popcorn. What is the mass of the bag of pears?

$$6 \times \frac{3}{10} = \frac{18}{10}$$

The mass of the bag of pears is $1\frac{8}{10}$ kilograms.

I read the problem. I read again.

As I reread, I think about what I can draw.

I draw a tape diagram to represent the problem. The first tape represents the mass of a bag of popcorn.

The second tape represents the mass of a bag of pears, which is 6 times the mass of the bag of popcorn.

The total mass of the bag of pears is unknown.

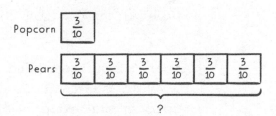

I can multiply $\frac{3}{10}$ by 6 to find the mass of the bag of pears.

$$6 \times \frac{3}{10} = \frac{6 \times 3}{10} = \frac{18}{10}$$

I rename the fraction as a mixed number, $1\frac{8}{10}$, because it makes sense to talk about mass as a mixed number.

REMEMBER

Divide. You may draw a model to help you.

2. 8,340 ÷ 3 = ___2,780___

Sample:

```
              0
             8 0
           7 0 0
         2,0 0 0
      3 ) 8,3 4 0
        - 6 0 0 0
          2 3 4 0
        - 2 1 0 0
            2 4 0
          - 2 4 0
                0
```

I see that 8,000 is the largest unit in the total. I can make 3 groups with 2,000 in each group. I write 2,000 above the total as the first partial quotient. I know that 3 × 2,000 is 6,000. I subtract 6,000 from the total.

```
          2,0 0 0
      3 ) 8,3 4 0
        - 6 0 0 0
          2 3 4 0
```

There is 2,340 remaining. I can make 3 groups with 700 in each group. I write 700 above the total as the next partial quotient. I know that 3 × 700 is 2,100. I subtract 2,100 from the total.

```
            7 0 0
          2,0 0 0
      3 ) 8,3 4 0
        - 6 0 0 0
          2 3 4 0
        - 2 1 0 0
            2 4 0
```

There is 240 remaining. I can make 3 groups with 80 in each group. I write 80 above the total as the next partial quotient. I know that 3 × 80 is 240. I subtract 240 from the total. There is nothing left to divide.

I add the partial quotients to find the quotient of 8,340 ÷ 3.

$$2,000 + 700 + 80 = 2,780$$

Copyright © Great Minds PBC

33

Name _____ Date _____

Use the Read–Draw–Write process to solve each problem.

1. A pancake recipe requires $\frac{3}{4}$ teaspoons of vanilla. A cake recipe requires 3 times as much vanilla as a pancake recipe. How many teaspoons of vanilla are required for the cake recipe?

2. It rains $\frac{7}{8}$ inches in Chicago. It rains 5 times that amount in Seattle. How many inches does it rain in Seattle?

REMEMBER

3. Divide. You may draw a model to help you.

$9{,}452 \div 4 =$ _____

$$4 \overline{)\,9{,}4\,5\,2}$$

Copyright © Great Minds PBC

Name _____   Date _____

34

1. Multiply by using the distributive property.

$5 \times 3\frac{2}{6}$

| 3 | $\frac{2}{6}$ |

| 3 | $\frac{2}{6}$ |

| 3 | $\frac{2}{6}$ |

| 3 | $\frac{2}{6}$ |

| 3 | $\frac{2}{6}$ |

$$5 \times 3\frac{2}{6} = 5 \times \left(\underline{\quad 3 \quad} + \frac{2}{6} \right)$$

$$= \left(5 \times \underline{\quad 3 \quad} \right) + \left(5 \times \frac{2}{6} \right)$$

$$= \underline{\quad 15 \quad} + \frac{10}{6}$$

$$= \underline{\quad 15 \quad} + \underline{\quad 1\frac{4}{6} \quad}$$

$$= \underline{\quad 16\frac{4}{6} \quad}$$

The mixed number, $3\frac{2}{6}$, is shown in the tape diagram as 2 parts, 3 and $\frac{2}{6}$. There are 5 tapes to show multiplying by 5.

$$5 \times 3\frac{2}{6} = 5 \times \left(3 + \frac{2}{6} \right)$$

I apply the distributive property and multiply each part of the mixed number by 5.

$$(5 \times 3) + \left(5 \times \frac{2}{6} \right) = 15 + \frac{10}{6}$$

I rename the fraction greater than 1 as a mixed number.

$$\frac{10}{6} = 1\frac{4}{6}$$

Lastly, I add the products.

$$15 + 1\frac{4}{6} = 16\frac{4}{6}$$

2. Multiply by using the distributive property. You may draw a model to help you.

$8 \times 4\frac{3}{5}$

$$8 \times 4\frac{3}{5} = 8 \times \left(4 + \frac{3}{5}\right)$$

$$= (8 \times 4) + \left(8 \times \frac{3}{5}\right)$$

$$= 32 + \frac{24}{5}$$

$$= 32 + 4\frac{4}{5}$$

$$= 36\frac{4}{5}$$

I decompose the mixed number into a whole number and a fraction.

Then I apply the distributive property and multiply each part of the mixed number by 8.

$$(8 \times 4) + \left(8 \times \frac{3}{5}\right) = 32 + \frac{24}{5}$$

I rename the fraction greater than 1 as a mixed number.

$$\frac{24}{5} = 4\frac{4}{5}$$

Lastly, I add the products.

$$32 + 4\frac{4}{5} = 36\frac{4}{5}$$

Copyright © Great Minds PBC

REMEMBER

3. Fill in the blanks to complete the two comparison statements that represent the equation $30 = 5 \times 6$.

   <u> 30 </u> is <u> 5 </u> times as much as <u> 6 </u> .

   <u> 30 </u> is <u> 6 </u> times as much as <u> 5 </u> .

I can think about $30 = 5 \times 6$ as representing two comparisons.

This tape diagram shows that 30 is 5 times as much as 6. I see a unit of 6, five times.

This tape diagram shows that 30 is 6 times as much as 5. I see a unit of 5, six times.

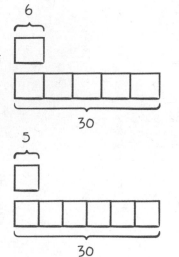

4. Write an equation to represent that 32 is 4 times as many as 8.

$32 = 4 \times 8$

I think about the statement 32 is 4 **times as many as** 8.

I write the equation 32 = 4 x 8.

34

Name _____ Date _____

Multiply by using the distributive property.

1. $3 \times 4\frac{2}{10}$

4	$\frac{2}{10}$

4	$\frac{2}{10}$

4	$\frac{2}{10}$

$$3 \times 4\frac{2}{10} = 3 \times \left(4 + \frac{2}{10}\right)$$

$$= (3 \times \underline{\quad}) + \left(3 \times \frac{\square}{10}\right)$$

$$= \underline{\quad} + \frac{\square}{10}$$

$$= \underline{\qquad}$$

2. $2 \times 5\frac{4}{12}$

5	$\frac{4}{12}$

5	$\frac{4}{12}$

$$2 \times 5\frac{4}{12} = 2 \times \left(\underline{\quad} + \frac{\square}{12}\right)$$

$$= (\underline{\quad} \times 5) + \left(\underline{\quad} \times \frac{4}{12}\right)$$

$$= \underline{\quad} + \frac{8}{12}$$

$$= \underline{\qquad}$$

3. Multiply by using the distributive property. You may draw a model to help you.

$4 \times 2\frac{3}{8}$

REMEMBER

4. Fill in the blanks to complete the two comparison statements that represent the equation $40 = 5 \times 8$.

    _____ is _____ times as much as _____ .

    _____ is _____ times as much as _____ .

5. Write an equation to represent that 54 is 6 times as many as 9.

Copyright © Great Minds PBC

Acknowledgments

Kelly Alsup, Lisa Babcock, Adam Baker, Reshma P. Bell, Joseph T. Brennan, Leah Childers, Mary Christensen-Cooper, Jill Diniz, Janice Fan, Scott Farrar, Krysta Gibbs, Torrie K. Guzzetta, Kimberly Hager, Eddie Hampton, Andrea Hart, Rachel Hylton, Travis Jones, Liz Krisher, Courtney Lowe, Bobbe Maier, Ben McCarty, Ashley Meyer, Bruce Myers, Marya Myers, Victoria Peacock, Maximilian Peiler-Burrows, Marlene Pineda, Elizabeth Re, Jade Sanders, Deborah Schluben, Colleen Sheeron-Laurie, Jessica Sims, Tara Stewart, Mary Swanson, James Tanton, Julia Tessler, Jillian Utley, Saffron VanGalder, Rafael Velez, Jackie Wolford, Jim Wright, Jill Zintsmaster

Trevor Barnes, Brianna Bemel, Adam Cardais, Christina Cooper, Natasha Curtis, Jessica Dahl, Brandon Dawley, Delsena Draper, Sandy Engelman, Tamara Estrada, Soudea Forbes, Jen Forbus, Reba Frederics, Liz Gabbard, Diana Ghazzawi, Lisa Giddens-White, Laurie Gonsoulin, Nathan Hall, Cassie Hart, Marcela Hernandez, Rachel Hirsh, Abbi Hoerst, Libby Howard, Amy Kanjuka, Ashley Kelley, Lisa King, Sarah Kopec, Drew Krepp, Crystal Love, Maya Márquez, Siena Mazero, Cindy Medici, Ivonne Mercado, Sandra Mercado, Brian Methe, Patricia Mickelberry, Mary-Lise Nazaire, Corinne Newbegin, Max Oosterbaan, Tamara Otto, Christine Palmtag, Andy Peterson, Lizette Porras, Karen Rollhauser, Neela Roy, Gina Schenck, Amy Schoon, Aaron Shields, Leigh Sterten, Mary Sudul, Lisa Sweeney, Samuel Weyand, Dave White, Charmaine Whitman, Nicole Williams, Glenda Wisenburn-Burke, Howard Yaffe

Credits

For a complete list of credits, visit http://eurmath.link/media-credits

OW 06.11.2021 1411